MAKING A SPLASH

How Humans Consume, Control and Care for Water

Colleen Nelson

Illustrated by Sophie Dubé

ORCA BOOK PUBLISHERS

Text copyright © Colleen Nelson 2025
Illustrations copyright © Sophie Dubé 2025

Published in Canada and the United States in 2025 by Orca Book Publishers.
orcabook.com

All rights are reserved, including those for text and data mining, AI training and similar technologies. No part of this publication may be reproduced or transmitted in any form or by any means, electronic or mechanical, including photocopying, recording or by any information storage and retrieval system now known or to be invented, without permission in writing from the publisher. The publisher expressly prohibits the use of this work in connection with the development of any software program, including, without limitation, training a machine-learning or generative artificial intelligence (AI) system.

Library and Archives Canada Cataloguing in Publication

Title: Making a splash : how humans consume, control and care for water / Colleen Nelson ; illustrated by Sophie Dubé.
Names: Nelson, Colleen, author. | Dubé, Sophie, illustrator.
Series: Orca timeline ; 9.
Description: Series statement: Orca timeline ; 9 | Includes bibliographical references and index.
Identifiers: Canadiana (print) 20240393236 | Canadiana (ebook) 20240393244 |
ISBN 9781459838697 (hardcover) | ISBN 9781459838703 (PDF) | ISBN 9781459838710 (EPUB)
Subjects: LCSH: Water—Juvenile literature. | LCSH: Water consumption—Juvenile literature. |
LCSH: Water use—Juvenile literature. | LCSH: Water conservation—Juvenile literature. | LCGFT: Informational works.
Classification: LCC TD348 .N45 2025 | DDC j363.6/1—dc23

Library of Congress Control Number: 2024941164

Summary: Part of the nonfiction Orca Timeline series for middle-grade readers, this illustrated book explores our relationship with water and how we use, control and take care of it.

Orca Book Publishers is committed to reducing the consumption of nonrenewable resources in the production of our books. We make every effort to use materials that support a sustainable future.

Orca Book Publishers gratefully acknowledges the support for its publishing programs provided by the following agencies: the Government of Canada, the Canada Council for the Arts and the Province of British Columbia through the BC Arts Council and the Book Publishing Tax Credit.

The author and publisher have made every effort to ensure that the information in this book was correct at the time of publication. The author and publisher do not assume any liability for any loss, damage, or disruption caused by errors or omissions. Every effort has been made to trace copyright holders and to obtain their permission for the use of copyrighted material. The publisher apologizes for any errors or omissions and would be grateful if notified of any corrections that should be incorporated in future reprints or editions of this book.

Cover and interior artwork by Sophie Dubé.
Design by Dahlia Yuen.
Edited by Kirstie Hudson.

Printed and bound in South Korea.

28 27 26 25 • 1 2 3 4

For Alex McGavin

There's no better way to cool off on a hot day than with a swim in a lake, ocean, river or pool. Did you know that people have been swimming since 8000 BCE? Archaeologists found the evidence in cave paintings in the western part of the Sahara desert.
ZORANM/GETTY IMAGES

CONTENTS

INTRODUCTION 1

ONE 5
Settlements

TWO 13
Agriculture

THREE 21
Ingenuity

FOUR 29
Water Security

FIVE 39
Spirituality

SIX 47
Exploration

SEVEN 55
Currents and the Climate Crisis

EIGHT 63
Equity

NINE 71
Wetlands and Water Bodies

GLOSSARY 81
RESOURCES 83
ACKNOWLEDGMENTS 84
INDEX 85

INTRODUCTION

What is your favorite memory of water? Is it splashing in a puddle? Bath time? Jumping in a lake or fishing off a dock? Maybe paddling in a canoe or running through a sprinkler? Like all living things, humans depend on water for many things, from drinking and bathing to growing food and getting around.

As a child, I spent summers at our family cottage on Falcon Lake, Manitoba. My days were all about hanging out on the dock, swimming and canoeing. I grew up chasing minnows and eating fish caught from the lake. Beavers, bald eagles and loons were common sights at the lake. Once a ferocious spring thaw mangled docks and destroyed boathouses, reminding us of water's power. There were days when the lake was so calm a loon's call could be heard for miles. On other days, we bobbed in a roller coaster of white-capped waves as they crashed on the shore. We had a pump house that brought water into the cottage for washing and flushing the toilet, but for drinkable water we filled up jugs at a nearby well. The water from the well was crisp, cool and tasty. Back then, despite being surrounded by water, or maybe *because* I was surrounded by water, I took it for granted. It was abundant and accessible where I lived. Looking back, I was privileged to have had my childhood soaked with happy memories.

As you read this book, think about your relationship with water. How has it shaped your life? How does it affect you now? Most important, what can you do to protect it?

I asked myself these questions as I wrote this book and was inspired to make some changes. I set up rain barrels to collect water for my garden, fixed a leaky faucet and cut out plastic waste wherever I could. I also switched to biodegradable dish soap and laundry detergent. While these changes might be small, they are important. Anything we can do to care for the world's water is significant—and we can all do something!

According to the United States Environmental Protection Agency (EPA), the average 1,000-square-foot roof can collect hundreds of gallons of rainwater each year using a rain barrel.
IMGORTHAND/GETTY IMAGES

WHERE DID WATER COME FROM?

How did water get here? Four and a half billion years ago, the earth was too hot for liquid water to survive on its surface. Scientists believe water may have arrived by asteroid after the planet cooled or was released as water vapor from the earth's core. Either way, in all that time the amount of water on our planet hasn't changed.

The amount of water on Earth has never changed, but its form has. Water is constantly moving through the four stages of the water cycle: evaporation, condensation, precipitation and collection. You might be drinking the same water the dinosaurs did!
HARVEPINO/SHUTTERSTOCK.COM

800 Port cities and trade routes

ONE

SETTLEMENTS
HOW WATER DETERMINES WHERE WE LIVE

If you were to look at a map of the places where most people live, you'd discover they are near water, whether that be an ocean, a river or a stream. In fact, most of the world's largest cities are located on a waterfront. The first settlements grew where people had easy access to fresh water for farming and domestic use, like drinking and cooking. Rivers and oceans made trade and travel easier, and over time bustling economies developed and populations grew. Tokyo is a perfect example. It started as a fishing village in a protected bay of the Pacific Ocean and has grown to become the largest city in the world, with a population of more than 37 million people.

YAORUSHENG/GETTY IMAGES

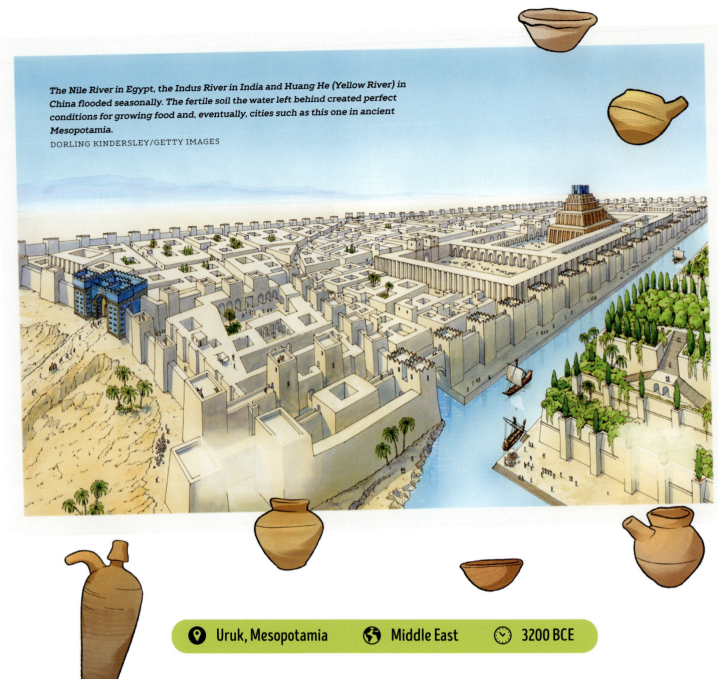

The Nile River in Egypt, the Indus River in India and Huang He (Yellow River) in China flooded seasonally. The fertile soil the water left behind created perfect conditions for growing food and, eventually, cities such as this one in ancient Mesopotamia.
DORLING KINDERSLEY/GETTY IMAGES

📍 Uruk, Mesopotamia   🌍 Middle East   🕒 3200 BCE

FROM MUD TO METROPOLIS

Archaeologists believe that agriculture began 30,000 years ago in Mesopotamia, thanks to its proximity to the Tigris and Euphrates Rivers. Soon after humans learned to farm, the world's first city, Uruk, was established. Why did people settle here? What made this place so special? The answer is mud.

The rivers flooded every year, leaving rich, fertile soil behind. The predictable seasons and climate made the location perfect for growing crops. Farmers set up permanent shelters to tend to their crops, and towns developed. The reliable food source meant one person could farm while another could specialize in a trade, like making pottery or grinding grain into flour. Over time, the *standard of living* improved, and the population grew. At its height, Uruk had temples, neighborhoods and defensive walls and was a mud-brick *metropolis* of 50,000 people.

| Aden, Yemen | Middle East | 🕒 800 |

Aden is one of the largest natural ports in the world. It's important as a refueling station for sea tankers. Recent conflicts have destabilized the region, leading to a decrease in the amount of shipping traffic using the port.
SANTIAGO URQUIJO/GETTY IMAGES

SILK AND SPICE

As people became better shipbuilders and sailors, a booming trade route developed in the Indian Ocean. People were eager for spices, silk and gunpowder from China and gold, textiles and jewels from Africa and India. Natural harbors where boats could safely dock became a hub for the exchange of goods. One of those stops was Aden in Yemen. Its economy and population grew as people moved there to take advantage of the opportunities that came with trade. Other cities, like Guangzhou in China and Zanzibar in Tanzania, also grew into bustling cities with thriving markets. It wasn't just goods being exchanged either. Trade helped spread culture, ideas and technological innovations too.

Moving goods by boat is still one of the most cost-effective methods of transportation. Cities on major waterways have strong economies and lots of job opportunities. It's not surprising that, like Aden, many of the world's biggest cities also have ports. The Port of Shanghai is one of the busiest, and Shanghai is also one of the world's most populous cities.

IMMIGRATION HUB

Some settlements grew because they were located on a trade route, but New York City grew because it was the first stop for people arriving by boat. Between 1892 and 1954, the doors to the United States were opened to people from around the world. Twelve million immigrants arrived by boat and passed through Ellis Island, the immigration center in New York Harbor. Most came from Europe and were fleeing war, famine or economic instability. Some moved on to other parts of the United States, but many stayed in New York City and found a home in ethnic neighborhoods around the growing metropolis. To this day, New York remains one of the most ethnically diverse places in the world. According to *World Atlas*, there are almost 20 million people living in New York City, and more than 800 languages are spoken!

The Statue of Liberty stands on Liberty Island, very close to Ellis Island. The statue was designed to represent freedom and enlightenment and was a welcome sight for new immigrants fleeing difficult conditions.

 Mississippi River, US   North America   1803

Paddle wheelers, or Mississippi steamboats, are unlike boats used anywhere else in the world and are considered a cultural icon in this region of the United States.
EDWIN REMSBERG/GETTY IMAGES

RIVER CITIES

Around the world in their traditional homelands, Indigenous Peoples were the first to make settlements near waterways. Not only were these locations good for trade and transportation, but they helped with defense. My hometown, Winnipeg, Manitoba, grew on the banks of the Red and Assiniboine Rivers, a site used for seasonal Indigenous settlements 6,000 years ago. Many European capital cities like London, Paris, Rome and Budapest also grew in importance thanks to their ideal riverbank locations. During the Middle Ages, castles were built on waterways so attackers could be spotted as they sailed into view. Centuries after the rise of European cities, North American settlers also saw the advantages of building cities on rivers.

THE MIGHTY MISSISSIPPI

When President Thomas Jefferson bought the Louisiana Territory, land stolen from Indigenous Peoples, from Napoleon in 1803, he doubled the size of the United States and got full access to an important waterway—the Mississippi River. Steamboats chugged up and down the river, moving goods and people, and cities like Minneapolis, St. Louis and New Orleans grew along its banks.

The *Industrial Revolution* ushered in a new era for cities on rivers. Manufacturing and transportation became more efficient. Cities like New Orleans, for example, became a hub for the cotton industry. Unfortunately, another industry also grew thanks to the Mississippi River—the slave trade. Plantation owners in the south relied on enslaved people to work the land and the mills. The Mississippi River acted as a thoroughfare to move enslaved people from upriver slave markets to plantations in the south.

It can take a long time for floodwater to recede. All that water must seep through rock layers into groundwater aquifers before the land can dry.
MICHAEL HALL/GETTY IMAGES

📍 Dhaka, Bangladesh
🌐 Asia   🕐 2050

WASHED AWAY

While water is often a reason to settle in a place, it can also push people away. Villages along the banks of the Mehgna River in Bangladesh used to be crowded with shops, homes, market stalls and farms. Today many of those places have been swept away by floodwaters. Rising sea levels, eroding riverbanks, erratic storms and creeping salt water have forced people to leave their ancestral homes for larger cities. A World Bank report predicts that by 2050, water will turn 140 million people into *internally displaced persons*. This includes large parts of the United States, coastal cities in Asia due to storm surges, and sub-Saharan African countries, which are vulnerable to flash floods because of poor *infrastructure*.

Busan's plan for a floating city features platforms for research, residential living and guestrooms. The residential buildings include a community backyard in the center of the platform.
OCEANIX/BIG-BJARKE INGELS GROUP

FLOATING CITY

Lots of cities develop near water, but Busan, South Korea, plans to build the world's first city *on* water. The proposed floating city is backed by the United Nations as a promising solution to the threat 570 cities around the world face from flooding due to rising sea levels. It will accommodate up to 10,000 people and have public squares, schools and markets. Most important, it will be designed to survive Category 5 hurricanes, which are the most destructive. Floating cities like this could be the solution to the problem of how to keep people safe in a world with a rapidly changing climate.

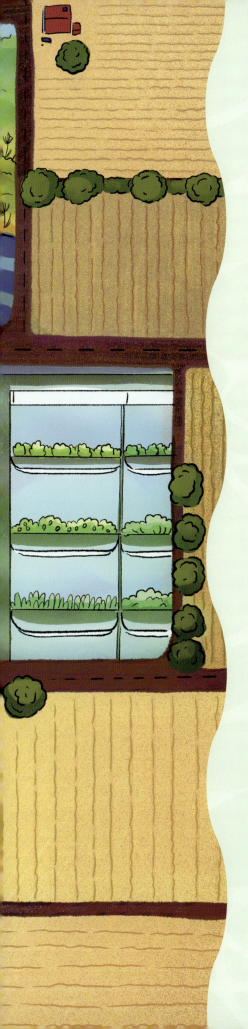

TWO

AGRICULTURE
THE WATER WE EAT

No matter what type of diet you have or where you live, you rely on farmers to feed you. Farmers, in turn, rely on soil, sunlight and water to grow their crops. A lot of available fresh water—70 percent!—is used for agriculture. Did you know it takes 275 gallons (1,250 liters) of water to grow 2.2 pounds (1 kilogram) of lentils? To produce the same amount of beef, 2,860 gallons (13,000 liters) of water is needed!

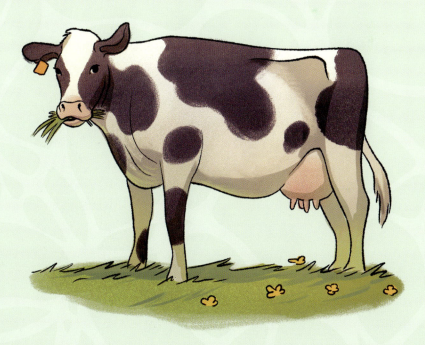

We actually "eat" more water than we drink. These hidden amounts of water contribute to our *water footprint,* which is the amount of water we use in our daily life. As human civilization has changed, so have agricultural practices.

Twelve thousand years ago, farmers relied on hand tools to water and irrigate the fields—much different from the large and expensive *irrigation* machinery some commercial farms use today.

Plants need water to help with photosynthesis, the process of turning sunlight into energy, and to move nutrients from the soil into the stem and leaves.
DUSAN STANKOVIC/GETTY IMAGES

A shaduf was a water-lifting device that appeared sometime after 1500 BCE. Using a bucket and a counterweight on opposite ends of a long pole, farmers lowered the bucket into the river, then poured the water into irrigation channels cut into the land.
THOMAS FAULL/GETTY IMAGES

 Tanis, Egypt   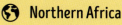 Northern Africa   1500 BCE

EARLY FARMERS

Ancient Egyptian civilization began on the banks of the Nile River. Seasonal floods left behind fertile soil in the **delta**, the low-lying, fan-shaped area at the mouth of the river, which was perfect for growing crops like barley, wheat and flax. The Egyptians were the first people to use irrigation, a method of bringing water to the crops. Basins were dug near the river to collect floodwater, and a network of channels extended out in a grid pattern. Farmers released water from the basin, and it flowed through the channels into the fields.

The Egyptians based their seasons around the Nile. They kept careful records of the Nile's height using a device called a nilometer. These innovations allowed the Egyptian farmers to feed cities and created one of the most stable and longest-lasting civilizations of all time.

A NEW "OLD" TECHNIQUE

Seven hundred years ago, Aztec farmers invented floating islands for growing food. Called chinampas, they are artificially constructed rectangular strips of land built on shallow lakes. Residents of Mexico City were reacquainted with the value of the centuries-old farming practice during the COVID-19 pandemic, when supply-chain issues and the closure of public markets meant people had to source food elsewhere. People once again turned to chinampas for locally grown food based on this traditional and sustainable farming practice.

The rice terraces of the Philippine Cordilleras are designated a UNESCO (United Nations Educational, Scientific and Cultural Organization) World Heritage Site, which means they have been determined to be of outstanding universal value.
R.M. NUNES/GETTY IMAGES

Philippine Cordilleras, Philippines | **Asia** | **1st Century**

TERRACED FARMING

Farming with basic tools is a challenge on flat ground. Imagine doing it on steep mountain slopes where water runs downhill. The Ifugao People of the Philippine Cordilleras are famous for the way they cut steps, or terraces, into the hillsides to trap rainfall and prevent it from washing away soil. The terraces are especially impressive because of the area's high altitude and the way the community works together to manage the complex system of dams, channels and bamboo pipes.

Terraced farming has also been used by the Inca in the Andes Mountains of South America. The Inca engineered their way past steep slopes and a harsh climate to grow potatoes, corn and quinoa. At the height of Inca civilization, the system of terraces covered one million hectares (10,000 square kilometers) of land throughout Peru.

Saskatchewan, Canada | North America | 1929

WHAT HAPPENS IN A DROUGHT?

Farmers are at the mercy of Mother Nature when it comes to precipitation. Throughout history there have been droughts—periods of very little (or no) precipitation. The *groundwater* dries up and farmers can't water their crops.

My grandfather grew up during the Dirty Thirties, when a drought swept the Canadian prairies. His family farm survived because his father had dug a small retention pond on their property, which his mother used to water a vegetable garden and feed their animals. As the landscape went from lush fields of green to a sandy desert, many farmers were forced to abandon their farms.

This drought was made worse because farmers had cleared large areas of land, dug up trees and filled in *reservoirs* to grow wheat. When the drought ended, important lessons had been learned. Trees were planted to create shelterbelts for preventing soil from drifting and trapping snow, areas were regrassed, and wetlands were restored. Farmers also dug irrigation channels and set up water-pumping stations in their fields so they'd be prepared for times when the rain didn't come.

HOW FARMERS WATER THEIR CROPS

Farmers dig wells by hand or with heavy equipment to access groundwater, which is brought to the surface by a pump.

Spray irrigation, similar to the kind of sprinkler system many people use to water their lawns, is used on large farms. Unfortunately, much of the water evaporates or is blown away before hitting the crop.

Some farmers dig wetlands to catch irrigation runoff or rainwater for use during dry spells. The wetlands also capture sediment, nutrients and chemicals so they don't leave the farm and pollute lakes and rivers.

Drip irrigation is a newer and more efficient water-use technique. Water runs through pipes with tiny holes in them to deliver water directly to the roots of the plant. Less water is used, and less is wasted due to evaporation.

SUSTAINABLE FARMING

Water shortages affect farmers all over the world. The United Nations says that by 2050, drought could affect 75 percent of the world's population. So how do we fight it? Ibrahim Thiaw, executive secretary of the UN Convention to Combat Desertification, suggests that we need to restore land, mimicking the natural landscape as much as possible. Other strategies, such as drip irrigation and *agroforestry* (combining trees and shrubs with crops), have also been effective.

The canopy and root systems of trees in agroforestry improve soil quality and reduce erosion. They also provide shade and prevent water loss from evaporation. If you have a garden, consider planting trees nearby. Your plants will thank you.
HADYNYAH/GETTY IMAGES

HYDROPONIC FARMING

Thanks to chemistry, data science and engineering, the farms of the future might be in warehouses. Bowery Farming in New York City is an example of a sustainable, innovative method of *hydroponic* food production. Housed in a giant warehouse operating 365 days a year, plants are grown vertically under LED lights and fed water that is mixed with fertilizers and minerals. Issues like water loss, soil erosion and drought don't affect hydroponic farms. Best of all, because large-scale hydroponic farms recirculate water to minimize waste, they consume 90 percent less water than farms that use traditional crop-watering methods.

THREE

INGENUITY
MAKING WATER WORK FOR US

People have been harnessing water for centuries for use in everything from steam engines to hydro dams. The first experiments in using the power of motion in water led to modern hydroelectric dams that produce about 20 percent of the world's energy. In coal, gas, nuclear and other kinds of power plants, water is used to produce electricity—fuel is used to heat the water into steam, and generators turn the steam's energy into electricity.

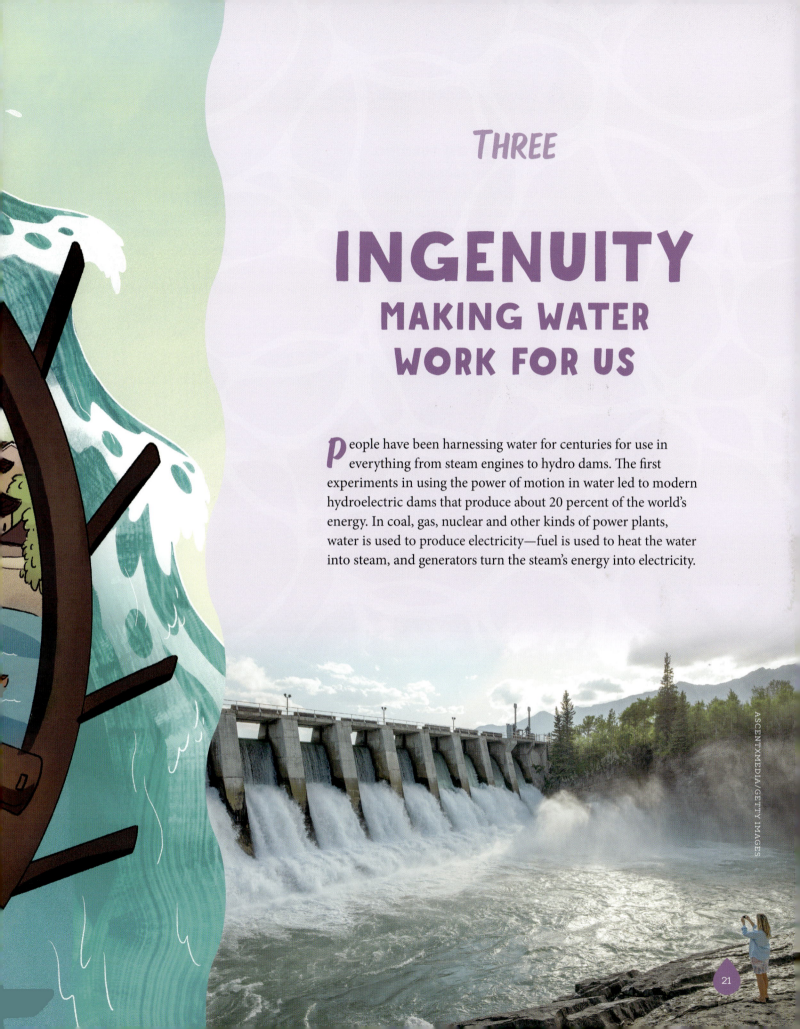

Barbegal Watermill Complex, Arles, France | Europe | 2nd Century

WHEELS KEEP ON TURNING

Roman engineer Vitruvius is believed to have invented the upright waterwheel, an ingenious device that used the power of moving water to do jobs like grinding grain, raising water for irrigation and felting cloth. The Barbegal watermill complex near Arles, France, produced enough flour to feed 27,000 people. The use of waterwheels spread through Europe during the Middle Ages. They were especially useful to landowners who'd lost workers due to illness or war. Watermills situated along babbling brooks and rushing rivers became familiar sights, and many have been there for centuries.

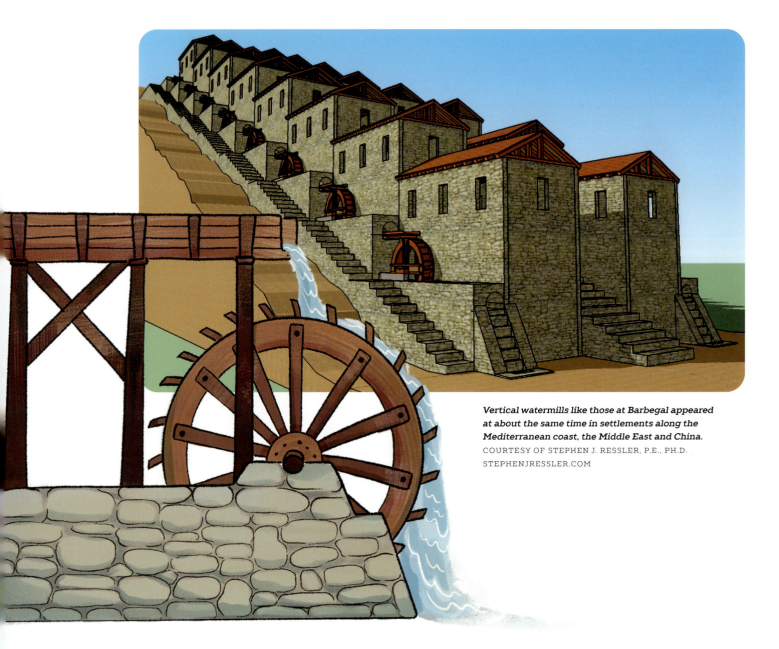

Vertical watermills like those at Barbegal appeared at about the same time in settlements along the Mediterranean coast, the Middle East and China.
COURTESY OF STEPHEN J. RESSLER, P.E., PH.D.
STEPHENJRESSLER.COM

Beijing, China
 Asia   605

IT'S NOT GREAT, IT'S GRAND!

The Great Wall wasn't China's only massive construction project. During the fifth century, Chinese leaders wanted a faster way to get the food grown in the south to the soldiers stationed in the north. The solution was to dig canals and let water do the work. By the time the Grand Canal was completed hundreds of years later, it stretched over 1,200 miles (over 1,700 kilometers) from the capital in the north (now known as Beijing) to the bustling port city of Hangzhou in the south, linking two of China's important rivers, the Huang He (Yellow) and Chang Jiang (Yangtze).

It took 5.5 million laborers to build the Grand Canal, and almost half of them died during the construction. The Grand Canal continues to be a busy and important transportation route, with 100,000 vessels moving millions of tons of goods through it each year.
SHUIGE/GETTY IMAGES

STEAM POWER

If you've ever kept a lid on a pot of water and seen it pop off as the water boils, you've witnessed steam power at work. A steam engine works on the same principle. Coal is used to heat water, and the steam created is forced into a smaller chamber to increase the pressure. As the steam exits, it pushes a piston and drives a wheel to create motion. The modern steam engine was invented in 1765 and was used to power ships, factories, trains and even cars. It is also credited with starting the Industrial Revolution, because it made production and transportation more efficient. In modern times, energy from nuclear fission (splitting of atoms) and geothermal (heat from Earth's crust) are used to boil water and create pressurized steam without producing greenhouse gases.

BEPPEVERGE/GETTY IMAGES

Agreements between developing countries to trade hydroelectric equipment and technology support their transition to hydroelectric power—a step toward reducing their reliance on fossil-fuel-emitting energy sources.
THIANCHAI SITTHIKONGSAK/GETTY IMAGES

 Appleton, Wisconsin, US   North America   1882

THE INVENTION OF HYDROPOWER

When the world's first hydroelectric plant began operating on the Fox River in Appleton, Wisconsin, in 1882, it produced enough electricity to power Henry James Rogers's home, the plant itself and a nearby building. Today the Three Gorges Dam in China is the world's largest hydroelectric dam, providing power to 60 million people.

WATER'S ENERGY

Hydroelectric power is created when fast-running or falling water flows over turbines, making them spin. The energy produced is collected and stored. To provide large amounts of hydroelectric power, dams are built to control the flow of water from high to low elevation. Unlike other sources of energy, hydropower doesn't require the burning of *fossil fuels*. Experts predict that by 2050 our energy needs will have increased by at least 25 percent. Many countries are setting goals to reduce the amount of carbon pollution they emit, which means there will be a growing need for *clean energy*, like hydropower.

Despite being a renewable source of power, hydroelectric power plants have their downsides. Flooding, ecosystem damage and pollution are all issues that environmental organizations have raised. Dam projects often force people, especially Indigenous Peoples, off their lands. During the construction of Three Gorges Dam, 1.3 million people needed to be relocated, and the dam's reservoir has been blamed for an increase in landslides and earthquakes in the area. The best solution to these problems is to use less power.

It takes a team of professionals to design a dam. Hydroelectric engineers design and build the dams while mechanical engineers oversee the pipework, valves and floodgates. Technical drawings are produced by civil engineers.
HISTORIC AMERICAN ENGINEERING RECORD (HAER), NPS/WIKIMEDIA COMMONS/PUBLIC DOMAIN

FROM PRISTINE TO POISONED

In 2015 construction on the Muskrat Falls Generating Station began in Labrador, Canada. The hydroelectric dam sounded promising. It would provide employment and power to the people in the Maritimes. But further investigation showed it posed a risk to Rigolet, an Innu community on the shores of Lake Melville. The people who live there rely on the river for their diet of brook trout, salmon and seal meat. Construction of the dam would contaminate the food chain with methylmercury, a toxin which is linked to severe illness, birth defects and even death. Despite scientific evidence proving the dangers, and protests from community members, the project continued, effectively ending a traditional way of life that has existed for hundreds of years.

FAUSTO CICILIOT/GETTY IMAGES

Hydropower dams often worsen conditions for the communities in the surrounding areas. Governments have the difficult task of finding a way to balance the need for renewable energy with protecting the land and its people.
ERIK.ALTITUDE/FLICKR.COM/CC BY-ND 2.0

📍 Orkney Islands, Scotland 🌍 Europe 🕐 2050

TURNING THE TIDES

An untapped power source might be lapping our shores. A growing number of companies are developing technology to harness tidal energy. Tidal energy refers to the predictable movement of water during tidal changes. A floating machine called a tidal stream generator captures energy from the moving water to turn the blades of a turbine.

Tidal energy has a lot of potential as a clean energy source. It occurs naturally and is more predictable than the wind or sun. The idea was first introduced by an English inventor in 1851, and the technology has advanced over the years. In 2011 the Sihwa Lake Tidal Power Plant opened in South Korea. It is the largest tidal installation in the world. It generates power twice a day at high tide as the water flows past ten turbines.

Currently tidal energy production is small, but this is starting to change, especially in places like the coastal waters of Scotland, where some of the world's strongest tidal *currents* can be found. The hope is that by 2050, tidal energy will account for 11 to 20 percent of the United Kingdom's electricity needs.

While harnessing tidal energy has a lot of benefits, there are also concerns about the high cost and the injuries that turbines inflict on marine animals.
IMAGINIMA/GETTY IMAGES

ANDRIY ONUFRIYENKO/GETTY IMAGES

WATER-POWERED CARS

In the 1970s inventor Stanley Meyer announced that he had invented a car that ran on water. His claim was proven to be fraudulent, but it hasn't stopped some manufacturers from working to make the dream a reality. While water can't combust (burn) like gasoline, its molecules can be separated in a process called electrolysis. The hydrogen molecules could power fuel cells in the car's engine. A water-powered vehicle would be ideal because water is clean and sustainable. The big drawback is that separating hydrogen requires a lot of energy, and until this obstacle is overcome, cars that use water for fuel will remain a dream for the future.

FOUR

WATER SECURITY

WHAT HAPPENS IF THE TAPS RUN DRY?

For many people, turning on a tap brings clean water, but for others that's not the case. Changes in climate, growing populations and water mismanagement have increased the number of people experiencing water insecurity. According to UN-Water, 1.42 billion people, including 450 million children, live in areas of high or extremely high *water vulnerability*. Water insecurity also drives conflict and migration as people leave their homes for a reliable water supply.

Because water is a limited resource, it is up to all of us to conserve it. Turning off the tap when you brush your teeth or wash your hands, taking (short) showers instead of baths and fixing household leaks might seem like insignificant actions, but they *do* matter. On a larger scale, governments need to reimagine how water is managed and protected to ensure water insecurity becomes a problem of the past, not the future.

El Castillo is a step pyramid in the center of the ancient Maya town of Chichen Itza. The town's name translates as "at the mouth of the well of the Itza." Access to water in cenotes, or naturally formed open wells, allowed the Maya civilization to flourish.
MATT CHAMPLIN/GETTY IMAGES

📍 **Yucatán Peninsula**   🌐 **North America**   🕐 **950**

MAYA MYSTERY

The fall of the ancient Maya civilization in Mexico has long been a mystery. Recent evidence from NASA's Earth Observatory suggests that water scarcity may have been the cause. By 950 the Maya had cut down most of the native forest around them and replaced it with agricultural crops to feed the growing population. Cutting down the forest didn't create a drought, but scientists believe it made naturally occurring droughts worse. Here's why. The Yucatán Peninsula's dark-green forests absorb light from the sun, allowing the warm air to rise and condense into rainfall. But when the Maya cut down the forest and replaced it with light-colored crops, the sun's light was bounced back into space, releasing less moisture into the air to make rain-making clouds. This resulted in a decline in precipitation. This chain of cause and effect that forms a circuit is called a ***feedback loop***. The effects wouldn't have been felt right away, but over decades less precipitation would have led to crop failures and wars over dwindling water resources. The Maya unknowingly created their own climate change. Twelve hundred years later, have humans learned from these past mistakes?

Saudi Arabia | Middle East | 2015

HOLD THE SALT!

With oceans covering 70 percent of Earth's surface, desalinating the seawater (removing salt from it) seems like an obvious way to end water shortages. Saudi Arabia's Ras Al-Khair Power and Desalination Plant has the capability to produce 274 million gallons (728 million liters) of water a day. There are drawbacks to desalination, however, mainly that it is expensive because it consumes a lot of energy.

HOW IT WORKS

Desalination separates the salt and water by heating the liquid and trapping the water vapor until it cools. This process occurs a few times to make the water drinkable. The other method is reverse osmosis desalination. The salt water is pushed through a membrane that allows only the lighter water molecules to pass through. Both desalination methods require a lot of energy. Engineers are constantly looking for ways to improve the process, or to use an alternate energy source like solar or wind to power the plants.

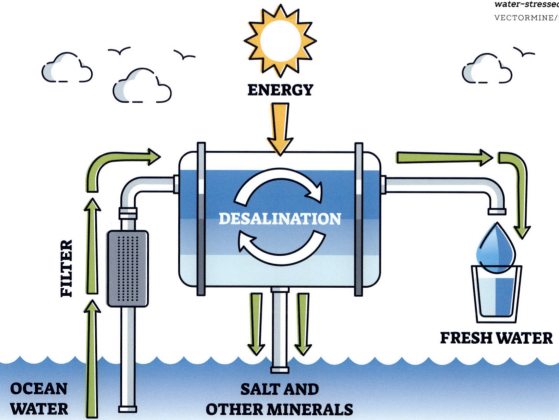

Researchers at MIT are working on a desalination device that uses solar power to produce clean, drinkable water. This kind of innovative thinking might provide solutions for water-stressed coastal countries.
VECTORMINE/SHUTTERSTOCK.COM

AREAS OF HIGH OR EXTREMELY HIGH
WATER VULNERABILITY

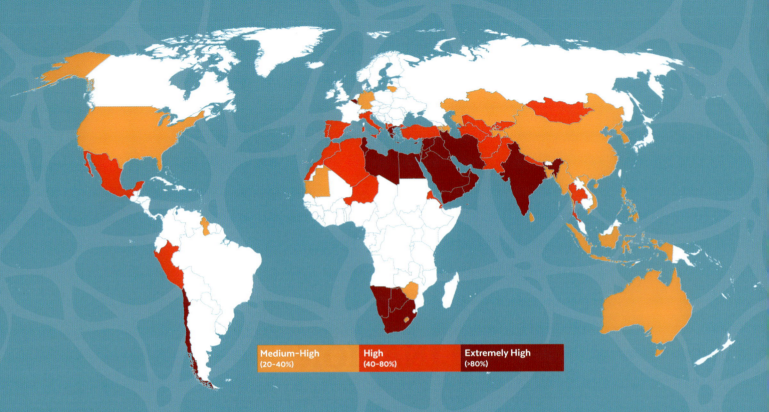

Medium-High (20-40%) | High (40-80%) | Extremely High (>80%)

Lack of precipitation, overconsumption, climate events and groundwater pollution all lead to water vulnerability. This map shows where in the world people are most at risk of water vulnerability.
SOURCE: WRI'S AQUEDUCT WATER RISK ATLAS, AQUEDUCT COUNTRY RANKINGS, 2019. WRI.ORG/AQUEDUCT

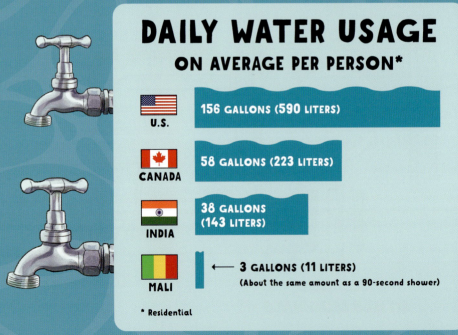

DAILY WATER USAGE
ON AVERAGE PER PERSON*

- **U.S.** — 156 GALLONS (590 LITERS)
- **CANADA** — 58 GALLONS (223 LITERS)
- **INDIA** — 38 GALLONS (143 LITERS)
- **MALI** — 3 GALLONS (11 LITERS) (About the same amount as a 90-second shower)

*Residential

SOURCES: CDC; STATISTICS CANADA; WATERMEDIA.ORG

Ninety-four percent of the population in high-income countries in Europe, as well as the United States and Canada, has access to safely managed water. In sub-Saharan Africa, that number drops to 31 percent.
PHILIPPE LISSAC / GODONG/GETTY IMAGES

📍 Burkina Faso 🌍 Africa 🕒 2015

LONG WALK FOR WATER

Some areas, like sub-Saharan Africa, are naturally drier and prone to seasonal droughts. Many rural communities don't have running water in their homes, or a village well, so the task of collecting water falls to women and girls. The job is risky, and girls are vulnerable. It's also time-consuming—which leaves no time for school—and difficult. Once the containers are filled, the walk home is even harder. Water is heavy!

One of the girls who made this daily journey was Georgie Badiel. One of 10 children, she would wake up every morning at six o'clock to start the three-hour journey to collect water with her grandmother and female cousins. Memories of her childhood walk for water stuck with her even after she grew up and became an international supermodel. She's used her influence to tackle the issue of water scarcity and gender inequality.

DIGGING WELLS AND CHANGING LIVES

The Georgie Badiel Foundation raises money to dig wells across Burkina Faso. The foundation also aims to empower women, so they are also taught how to maintain and repair the wells. Badiel's work has changed the lives of girls across her home country and raised awareness about the impact of water insecurity.

As Day Zero approached, Cape Town residents waited in long lines to collect water from the city's natural springs.
MARK FISHER/SHUTTERSTOCK.COM

Cape Town, South Africa   Africa   2018

DAY ZERO

For the four million residents of Cape Town, the countdown was on to Day Zero, July 9, 2018, the day their city would run out of water. Three years of dry winters meant the reservoirs that held the city's water were almost empty. Day Zero would mean that water reserves had shrunk to less than 13.5 percent of capacity, making them too low to deliver what the city required. Government officials implemented water restrictions to avoid Day Zero. Residents couldn't use water outdoors, which meant no swimming pools or watering the garden. There was wide-scale repairing of leaks, and toilets had to be flushed with *gray water*. The restrictions helped but not enough, so a new rule was introduced: people were limited to 11 gallons (50 liters) of water per day.

The average Cape Town resident used 132 gallons (600 liters) of water per day before the restrictions, so cutting that down to 50 liters was a big challenge. But people pulled together and Day Zero was warded off. Then rain finally came, and the population of Cape Town breathed a sigh of relief.

LESSONS IN CONSERVATION

Cape Town residents learned an important lesson as they approached Day Zero—reducing water use was achievable! Whether it was collecting rainwater to water plants or saving gray water to flush toilets, everyone could do something to conserve water. Even with Day Zero avoided, many people continued to be mindful of their water usage.

DANIEL LLAO CALVET/GETTY IMAGES

TWO-MINUTE SHOWERS

During the water crisis, Cape Town residents were asked to shorten their showers. South African musicians recorded two-minute shower songs for people to listen to while they showered. When the song ended, so did their showers, and each shower used only 4 gallons (18 liters) of water.

WHO OWNS WATER?

Water ownership is something that Canadian activist Maude Barlow has been fighting against for decades. She believes *no one* should own it because water is a human right. In Canada, all water is owned by the government. There are laws about how water can be used and who has priority during a water shortage. In the United States, things get trickier and vary from state to state. Surface water like lakes and streams is owned by the government, but groundwater may be publicly or privately owned, which is why the Michigan government sold access to an underground aquifer to food-and-beverage company Nestlé for just $200 a year. The company bottles 100,000 times the amount of water the average resident uses. They package it in plastic bottles and sell it for a healthy profit.

Harvesting rainwater before it becomes runoff is a great thing for everyone to do, not just people in water-stressed places. The water can be filtered for drinking or watering a garden. Rain barrels or larger cisterns can be installed under the roof's downspout to collect the most water.
WATERDOTORG/FLICKR.COM

 Mexico City, Mexico   North America   2030

FACING THE FUTURE

As populations grow, major cities like Tokyo, Moscow and London will face water shortages in the future. Some places, like Mexico City, with 22 million people, are already experiencing them. Mexico City's **aquifers** are emptying faster than they can be refilled. Even today the city is **water-stressed.** Many residents go weeks without access to water. What will happen in 2030, when the city's population is expected to hit 30 million?

RAINWATER AND RUNOFF

Part of the problem is that despite high rainfall, the city has so much concrete that a large portion (40 percent) of the water runs off and ends up in sewers instead of sinking into the ground to fill aquifers. A rainwater catchment program for homeowners provides some relief. Cisterns are installed on the roofs of homes and schools to capture and store rainwater. The creation of more green spaces and retention ponds helps ensure that rainwater isn't wasted. With conservation strategies and proper management of our most precious resource, the looming water crisis facing many cities can be averted.

BLUE-GREEN SPACES

Our cities are full of hard surfaces. Concrete roads, sidewalks, driveways, parking lots and playgrounds mean that rainwater and snowmelt run into sewers instead of being absorbed into the aquifers. Creating more green space in your neighborhood is a great way to improve the health of your watershed. Blue-green infrastructure, which includes rooftop gardens, permeable paving stones, rain gardens and urban forests, captures that runoff before it hits the sewer system.

WITTHAYA PRASONGSIN/GETTY IMAGES

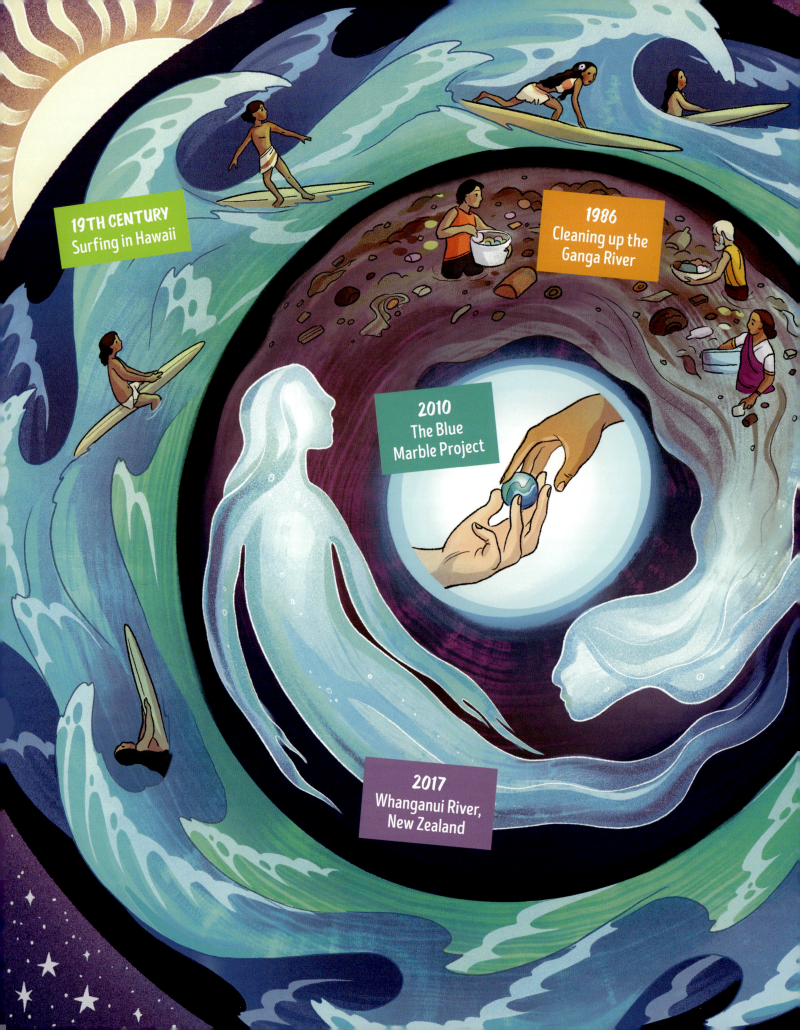

FIVE

SPIRITUALITY
FINDING CONNECTION THROUGH WATER

Water is a life-giving force. Without it, humans would not exist. For this reason, many cultures give water special significance, linking it to their beliefs through legends and myths. Rivers can be sacred, symbolic or part of ancestral origin stories. The spiritual connection between humans and water isn't just about worship. Quiet moments in blue space give us the chance to consider who we are and what our place on the planet might be—questions that are part of what makes us human.

RODRIGO FRISCIONE/GETTY IMAGES

Ancient Hawaiians referred to surfing as he'e nalu, which means "wave sliding," and would ask the kahuna, or priest, to pray for great waves.
WHITEMAY/GETTY IMAGES

SACRED WATERS

Some temples, like Kiyomizu-dera in Kyoto, Japan, are situated near water so visitors can meditate on the ever-changing beauty of nature. The temple was built in 778 after a monk discovered a waterfall on the slopes of Mount Otowa and had a vision that told him it would be a sacred place. Besides admiring the beautiful surroundings, temple visitors can drink from one of the three streams to give them success, longevity or love—but you can't drink from all three because that would be greedy.

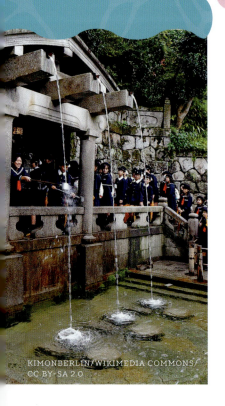
KIMONBERLIN/WIKIMEDIA COMMONS/ CC BY-SA 2.0

 Hawaii, US   Polynesia   19th century

PRAYING TO THE GODS

In traditional Hawaiian culture, surfing is more than riding the waves and hanging 10. Craftsmen prayed and made offerings to the gods before they shaped boards from sacred trees. Surfers prayed to Kanaloa, god of the sea, for success.

Missionaries who arrived at the island in the 1800s documented how surfing competitions were used to settle disputes. All Hawaiians owned a surfboard, but being a good surfer raised your status. The missionaries disapproved of the mixed-gender sport and the bare skin and gambling that accompanied it, and they tried to stop it. But surfing never went away. Today people travel from around the world to find the best surf spots in Hawaii.

MORE SPIRITUAL THAN SPORT

Just like the Hawaiians, other surfers view the activity not as a sport but as something spiritual. According to surfer and psychologist Timothy Leary, surfing is "just the individual dealing with the power of the ocean, which gets into the power of lunar pulls, and of tidal ebbs and flows." Researchers explain that when wave-riding, our ancestral connection to water plays a role. Some even claim that because life on Earth began in the oceans, surfing allows humans to tap into our origins and find solace in something greater than ourselves.

Ganges River, India Asia 1986

CAN THE GANGA BE SAVED?

The Ganga, or Ganges, River stretches about 1,500 miles (2,400 kilometers) across India and is one of the most spiritually significant bodies of water in the world. It is one of the seven sacred rivers in Hinduism. According to ancient Hindu scriptures, the river is said to be the physical embodiment of the goddess Ganga. Since the river is pure, Hindus believe that rituals performed in its water will bring fortune and wash away impurity. The river is a source of happiness and contentment. With a few sips, or a dip in the water, suffering is gone.

However, the river is also heavily polluted. Millions of gallons of raw, untreated sewage, as well as animal waste and toxic chemicals, flow into the river every day. The Ganga River has 3,000 times more dangerous bacteria and toxins than is suggested safe by the World Health Organization. Waterborne illness and superbugs (bacteria resistant to antibiotics) are linked to pollution. Despite the risks, 400 million people in India and Bangladesh continue to use the Ganga River for bathing, drinking and irrigation.

Various programs have been launched over the years, starting with the Ganga Action Plan in 1986, to clean up the river. Sadly, the actions have had little impact. The biggest obstacle to cleaning up the Ganga River is that many people believe it has the power to clean itself and doesn't need our help.

RELIGIOUS SIGNIFICANCE OF WATER

In Buddhism, water represents purity, clarity and calmness, and it is an important part of preparing to pray. At temples throughout Asia, scoops with bamboo handles sit in small pools of water. Before you enter the temple, you pour water over your hands to purify your hands and cleanse your heart.

Muslims must perform wudu, the washing of face, hands, arms, head and feet with water before prayer. Wudu cleanses worshippers of sins and symbolizes a fresh start and a purified state of mind.

Christians use holy water to bless themselves. During a baptism, water is sprinkled or poured on the head, or a person is immersed, to wash away sins and be symbolically reborn.

Jews also use water for a purification ritual in a mikveh, or bath, which must be connected to natural water.

Granting rivers and lakes personhood is a global movement. Indigenous people in such countries as Canada, Colombia and the United States hope their governments will follow New Zealand's lead and recognize the rights of nature.
MATTHEW MICAH WRIGHT/GETTY IMAGES

 Whanganui, New Zealand   **Oceania**   **2017**

ANCESTRAL RIVER RIGHTS

The Whanganui River is sacred to the Whanganui peoples, the Māori tribes also known as Ngāti Hau. They believe that the river was born when a teardrop fell from the Sky Father to the base of the tallest mountain. Located on New Zealand's North Island, the Whanganui Iwi have had an ancestral connection to the river that began long before settlers arrived in New Zealand.

But during the 1800s, Europeans arrived and altered the landscape with dams and dynamite. The fish and eels the Whanganui peoples had depended on for centuries were harmed, and pollutants were dumped in the river. Without consulting the Whanganui peoples, the settlers changed the river's natural flow so it could be used in a hydroelectric power dam. To the Whanganui peoples, they had desecrated not just a body of water but a sacred ancestor.

But how do you protect this water that is so much more than just a river? Declare it a person! That is what happened in 2017. The New Zealand parliament passed a law that gave the river the rights and powers of a legal person. The legislation signaled a shift away from the settler mindset—that land and its resources exist to be developed and used—and toward the Māori belief that land and water are their relatives and need to be respected as such. As the Whanganui peoples say, "I am the river, the river is me."

For the Bajau, a nomadic Indigenous People found in coastal areas of Southeast Asia, the connection to the ocean is an essential part of their culture. Their god of the sea, Omboh Dilaut, rules the ocean and the creatures within it.
KHAICHUIN SIM/GETTY IMAGES

Wherever You Are | The World | Today

LEAVE YOUR WATERMARK

What is your most powerful memory of water? That's the question that Mark Mattson, founder of Swim Drink Fish, wants you to answer. He believes that every person has a water story and that each one is unique. Your story can be spoken, written, filmed or illustrated and will be stored on the Watermark Project online. His goal is to collect 35 million stories from around the world. You can see an example of my watermark in the introduction of this book!

Mattson has been a water activist for a long time. His organization aims to restore and protect water so everyone can swim, drink and fish in it. Besides the Watermark Project, Swim Drink Fish has water-monitoring hubs to equip communities with the tools, skills and resources to monitor their water. They also support the Blue Flag award program to promote environmentally friendly beaches, boats and marinas. A recent project supported by the organization was to revitalize Toronto's 28-mile (46-kilometer) inaccessible lakefront with boat launches, lakeside pools and open-water swimming.

Research has proven that blue spaces have a profound impact on humans. Being around water boosts our emotional wellbeing. Just the sounds and sights of water capture our attention, freeing our minds and bringing comfort and belonging. If you haven't spent time in or near water recently, see if the research is right—does your state of mind improve when you're in a blue space?

CAVAN IMAGES/GETTY IMAGES

BLUE MARBLE PROJECT

To show his gratitude to the precious resource that water is, and to remind others of how connected we are through water, marine biologist Dr. Wallace J. Nichols began the Blue Marble Project. His mission was to pass a blue marble through every person's hand on Earth. The campaign is simple. Pass on a blue marble when you see someone doing good for our blue planet. That person, in turn, will pass it on to someone else, remembering that all actions are interconnected, both good and bad. Since it began in 2010, millions of people have received and passed on a blue marble in celebration of the idea that water is life.

SIX

EXPLORATION
RIVER ROUTES AND DISTANT SHORES

Early humans made their way to Australia by boat about 50,000 years ago on simple bamboo rafts. No one knows if they made it there on purpose or by accident. By 3000 BCE, Egyptians were traveling the Mediterranean to trade, and thousands of years later, Portuguese and Spanish explorers set sail looking for a trade route to China. Voyages were dangerous—illness, starvation and storms claimed the lives of many sailors. But curiosity, dreams of wealth and colonization compelled humans to continue exploring. These days our coastlines might be mapped, but that doesn't mean there is nothing left to discover.

L'Anse aux Meadows, Newfoundland, Canada   North America   1021

RAIDERS AND TRADERS

In their Northern European homeland, the Norsemen (or Vikings, as we know them now) were farmers, but to people living across the North Sea in other parts of Europe, they were fearsome raiders who invaded and pillaged coastal towns. They were also adventurers, setting off on arduous journeys in their distinctive longboats. These early explorers didn't have maps or compasses. Historians believe they used their senses and paid attention to nature. For example, if they saw birds, it meant land was nearby. They also used sun compasses to navigate and stay on course.

What's more incredible about the Viking journeys was the distance they covered. Archaeological evidence at L'Anse aux Meadows in Newfoundland proves that Vikings were the first European travelers to North America.

Viking longships had narrow hulls that withstood the punishing waves and storms of the open ocean.
GREMLIN/GETTY IMAGES

Seville, Spain · Europe · 1519

THE QUEST TO GO WEST

When Ferdinand Magellan set off with five ships and 270 men in 1519, his plan was to sail west until he got to Asia. If he succeeded, he'd be the first explorer to make the journey. Traveling by sea came with risks. Besides starvation, dehydration, illness and storms, there was also the danger of getting lost or being attacked. Plus not even Magellan knew if the route existed or, if it did, how long the journey would take. There were still those who believed the world was flat and worried Magellan would fall off the edge!

Magellan crossed the Atlantic and made it to South America. Then he led his ships around the tip of South America through the Strait of Magellan (named after him) and to the Sea of the South. By the time his crew had crossed the Pacific, they were starving, thirsty and reduced to eating the leather on their sailing equipment. When they landed in Guam, it was the first time they'd had fresh food in 99 days.

THE JOURNEY ENDS

Magellan made landfall on the island of Homonhon (in what is today known as the Philippines) with the goal of claiming the land for Spain. The explorer's journey ended when Magellan invaded nearby Mactan Island and was killed by the people who lived there. Without their leader, Magellan's remaining crew headed for home. Only one of Magellan's ships and 18 of his crew returned to Spain in 1522, but the voyage was considered a navigational success because he had proved that the world is round.

While Magellan is celebrated for being an explorer, he was also a colonizer intent on converting the Indigenous people he encountered to Christianity. Mactan chief Lapu-Lapu fought back. His people defeated Magellan's army, and the explorer died in battle.
DRMAKKOY/GETTY IMAGES

The Monument to the Discoveries in Lisbon, Portugal, recognizes the travels of early explorers like Magellan.
SYLVAIN SONNET/GETTY IMAGES

49

The Amazon River releases 53,000 gallons (200,000 liters) of fresh water into the ocean every second. That's almost 20 percent of all the fresh water that enters the sea!
FG TRADE/GETTY IMAGES

🔴 **Amazon River** 🌎 **South America** 🕒 **1542**

Orellana and his crew encountered a group of women who fought alongside the men, much like the fabled Amazonian women of Greek myths. Orellana likely named the river in their honor.
MAKASANA PHOTO/SHUTTERSTOCK.COM

GOING FOR GOLD

Francisco de Orellana didn't plan on being the first European to sail down the Amazon River. He hadn't traveled to South America to explore but to join in the bloody conquest of the Inca Empire. Wealth was his motivator, not curiosity.

ORELLANA'S JOURNEY

His captain, Francisco Pizarro, sent him ahead on an exploratory mission. Three days later it was time to return, but his crew refused. They wanted to continue down the river rather than battle the current upstream. With the threat of his crew abandoning him, Orellana agreed, but it wasn't an easy journey. Along the way they were attacked and threatened by tribes of Omagua and the Tapuya people, and they faced hunger and illness. When he finally reached the open sea, Orellana had led his crew to places no European had ever been. He'd traveled the entire 4,300 miles (6,920 kilometers) of the Amazon, the world's longest river.

MODERN EXPLORERS

Today's Amazon basin explorers aren't looking for conquests or wealth, and they aren't explorers by accident. Their journeys into the Amazon jungle are to research, document and photograph the river's complex ecosystem. While Orellana and other explorers intent on colonizing were often met with resistance from locals, explorers today respect, partner with and learn from Indigenous communities. The Amazon River is mapped out thanks to the efforts of early explorers, but with only a fraction of its plant and animal life identified, there is still much for modern-day explorers to discover about this **biodiverse** river ecosystem.

People who want to get up close and personal with the Amazon River can stay at eco-lodges deep in the jungle, where they are immersed in the local culture.
MATTHEW MICAH WRIGHT/GETTY IMAGES

Geothermal vents in the ocean floor deposit valuable metals like gold, silver, copper, cobalt and other rare elements in concentrations up to ten times greater than what is found on land.
GEOMAR BILDDATENBANK/WIKIMEDIA COMMONS/CC BY 4.0

JOHN R. PLATT, THE REVELATOR/FLICKR.COM/CC BY 2.0

HER DEEPNESS

Marine biologist and *oceanographer* Dr. Sylvia Earle has been exploring the ocean for seven decades. She's spent more than 7,000 hours underwater, on over 100 expeditions, in what she calls "the blue heart of the planet." As an ocean explorer, she's witnessed the human impact on marine ecosystems and has spent her life raising awareness and championing the importance of protecting and restoring them. In 2010 Earle and her team at Mission Blue, an organization she founded, designated the Gulf of California as the first Hope Spot, a place critical to the health of the ocean. Since then over 150 new Hope Spots have been added.

📍 **Clarion-Clipperton Zone**   🌐 **North Pacific Ocean**   🕒 **2030**

MAPPING THE SEA

With most of the world's coastlines mapped out, what is left to explore? According to UNESCO, a lot! The ocean covers 71 percent of our planet, but only 20 percent has been accurately mapped. Mapping the seafloor helps with ship navigation and climate modeling. Accurate measurement of the ocean's depth is important for scientists to understand the impacts of the climate crisis and learn about the creatures living in deep water. In 2017 the Seabed 2030 Project began, with the goal of using data from countries around the world to complete a map of the world's seafloor by 2030.

PROTECTING THE DEPTHS

With so little known about the ocean floor, efforts have been made to protect it. In 2023, 193 nations signed the High Seas Treaty, which aims to place 30 percent of the seas into protected areas by 2030. These areas will limit fishing, shipping and *deep-sea mining*. Environmental groups are concerned that mining disturbs animal breeding grounds, creates noise pollution and is toxic to marine life. One area specifically targeted for mining is the mineral-rich Clarion-Clipperton Zone between Hawaii and Mexico.

Littered along the seafloor are potato-sized chunks of manganese, nickel, copper and cobalt, minerals essential for *renewable energy* projects. While companies are eager to extract these deep-sea resources, environmental organizations are worried. With so little known about the depths of the ocean, harvesting the minerals will disrupt an ecosystem in which the majority of species are unknown to scientists.

DIVING THE DEPTHS

Tara Roberts is part of a team of Black divers, historians and marine archaeologists who search for the wrecks of ships that carried enslaved Africans across the Atlantic Ocean. Historians estimate that 12 million Africans were brought to the Americas, and anywhere from 500 to 1,000 ships may have wrecked on the way over. Only five have been found, and only two have been excavated and documented. Roberts and her team gather evidence, document and identify the sunken shipwrecks carrying enslaved people to bring information from these ships back to collective memory.

Victor Vescovo was the first person to dive in a *submersible* to the deepest points of the earth's five oceans. Part of his mission was to *sonar*-map the seafloor for Project Seabed 2030. Visiting the depths of the oceans led to some exciting discoveries, like unusual sea life in the Indian Ocean, and some disturbing ones, like the plastic bag and candy wrappers found at the bottom of the Pacific Ocean.

Jill Heinerth has led expeditions into icebergs, volcanic lava tubes and underwater caves. She's also created 3D maps of submerged caves to help us better understand cave systems and their hidden fresh water. "More people have been to the moon than to places that Jill Heinerth has explored deep inside our watery planet," said filmmaker and underwater explorer James Cameron.

VLVESCOVO/WIKIMEDIA COMMONS/CC BY-SA 4.0

JILL HEINERTH, INTO THE PLANET

SEVEN

CURRENTS AND THE CLIMATE CRISIS

CAN BALANCE BE RESTORED?

From record-breaking heat waves to deadly storms, the burning of fossil fuels like oil and coal has changed, and will continue to change, our climate. The heat trapped in the atmosphere by carbon dioxide is raising the temperature of the earth and the ocean. The warmer ocean temperatures affect ocean currents, disrupting local weather and global climate patterns. We see the impact as **glaciers** melt, sea levels rise and weather events like hurricanes and droughts get more frequent and severe. Honestly, it sounds scary, but the solutions to our climate crisis are out there. People know what to do to fix our planet, and we still have time to do it. "Knowledge is the superpower of the 21st century," says oceanographer Sylvia Earle. "Even the smartest people alive when I was born did not know what 10-year-olds today have available to them. That's truly cause for hope."

Bering Strait · **North America** · **40,000 years ago**

NATURAL CLIMATE CHANGE

The planet has experienced cold periods and warm periods for at least the last one million years. In 100,000-year cycles, global temperatures can change by 5.4 to 14.4°F (3 to 8°C). During the last ice age, which ended about 11,000 years ago, glaciers covered most of North America. With so much water locked in ice, the sea level fell, exposing a bridge of land, called the Bering Land Bridge, between Asia and North America. As Earth began to warm, glaciers melted and sea levels rose by as much as 52 feet (16 meters). Areas that were formerly connected, like Asia and North America, or the islands of Java, Borneo and Malaysia in Southeast Asia, were once again separated.

SLOWING THE WARMING

Fast-forward to today, when Earth is warming much faster. NASA's research shows the global temperature has risen by more than 2.45°F (1.4°C) since 1880. The good news is that according to the National Oceanic and Atmospheric Administration's recent findings, a global effort to lower carbon emissions will help slow down rising temperatures. To do this, developed countries need to phase out coal by 2030. The United Nations has even called for all countries to commit to *net-zero emissions* by 2050.

Ocean habitats like mangrove forests (mangroves are trees or shrubs that grow in areas flooded at high tide) and kelp forests are some of the most carbon-rich ecosystems on the planet.
MARIAKRAY/GETTY IMAGES

The white "bathtub ring" on Lake Mead's rocks is caused by calcium carbonate deposits left behind when the water level goes down. The bleached rock is a stark reminder of how much the water level has dropped.
BLOODUA/GETTY IMAGES

ANDREW PEACOCK/GETTY IMAGES

 Colorado River Basin and Mid-Atlantic Watershed
 North America   **2022**

WACKY WEATHER

Ocean currents are responsible for a weather phenomenon called La Niña, which caused people on both coasts of the United States to experience unusual weather in 2022. In the west, a water shortage was declared by the US federal government. An ongoing drought and rising temperatures had depleted the huge reservoirs of Lake Mead and Lake Powell, which provide water to 40 million people and millions of acres of farmland.

While there was too little water in one part of the country, the Mid-Atlantic *watershed* was getting too much. Over 11 days in 2022 in the northeastern United States, there were four flooding events that would normally be expected once every 1,000 years. These flash floods, or rapid rainfall causing flooding, resulted in loss of life and destruction of property. Streets turned into rivers, and people had to be evacuated from their homes.

HOW CLIMATE CYCLES WORK

La Niña and El Niño are climate patterns that happen in the Pacific Ocean and break the normal conditions. The complete system is known as the El Niño-Southern Oscillation (ENSO), and it is difficult to predict but can affect weather worldwide. During La Niña, the ocean cools down as more cold, deep water moves to the surface. *Trade winds* push warm water toward Asia, resulting in cool, wet weather in the east and warm, dry conditions in the west. When the trade winds weaken, El Niño occurs. Warm water is pushed to the west coast of the Americas and areas in the United States and Canada are drier and warmer than usual, but conditions are wetter in the US southeast and Gulf Coast.

El Niño and La Niña last nine to twelve months, or sometimes longer, and occur every two to seven years. If you think the effects of ENSO are getting more intense, you're right. As the climate crisis heats the ocean water, El Niños and La Niñas are having a bigger impact. The droughts in dry climates are more severe, and the flooding in wetter climates is more intense.

RAPIDLY DISAPPEARING

The Colorado River, which runs through the Grand Canyon, is famous for its turbulent rapids and is loved by white-water rafting enthusiasts. But things are changing. Hotter, drier weather, less snowpack in the Rockies and continued overuse by farmers and cities has lowered the water level of this great river. There is concern from some guides that they won't be able to raft the river anymore if the water dips much lower. Trees have been planted along the banks to provide shade, and upstream reservoirs have been released to raise water levels, but the message is clear: if we humans want to enjoy what nature has to offer, we must take care of it.

Who cleans up after a storm hits? When catastrophic damage has occurred, the government may declare a state of emergency for the affected area. Military personnel and additional support are called in to assist with road and building restoration, debris disposal and other cleanup tasks.
FRANKRAMSPOTT/GETTY IMAGES

 Atlantic Provinces, Canada   North America   2022

HURRICANES HIT ATLANTIC CANADA

The costliest hurricane season on record in Canada was in 2022. Hurricane Fiona, the strongest hurricane in Canadian history, hit the Atlantic provinces in late September. A few weeks later Hurricane Ian hit, causing storm surges and flooding.

Hurricanes are powerhouse weather events that form over the ocean in an area of low pressure called a tropical wave. Water must be at least 80°F (26.5°C) over a depth of 0.03 miles (50 meters) for a hurricane to develop. Scientists warn that as temperatures in the Atlantic Ocean and Gulf of Mexico rise, so does the risk of stronger hurricanes. Sometimes it's not just the hurricane winds that cause damage. A storm surge is an abnormal swell of water generated by the force of winds. The rush of water washes out roads and destroys homes. The storm surge that came with Hurricane Fiona killed three people and left thousands without power.

ENGINEERING SOLUTIONS

Solving the problem of changing water levels is something the Netherlands has been doing for centuries. Systems of dams and **dykes** defend low-lying cities such as Rotterdam against storm surges, and measures such as widening rivers, reinforcing coastlines with sand and building floating homes and warning systems have helped their flood preparedness.

FUTURE WEATHER EVENTS

A study published in the journal *Science Advances* claims that intense hurricanes and typhoons could double by 2050 in nearly all regions of the world. Some areas, like the Gulf of Mexico, which already sees a lot of hurricanes, will remain unchanged, while cities like Tokyo, Hong Kong and Honolulu will have a much higher probability of facing tropical storms.

Researchers hope that by sharing their findings, governments will take measures to make their cities less vulnerable to hurricanes and storm surges. Brett Sanders, a professor of civil and environmental engineering at the University of California, Irvine, advises that cities map and monitor flood risk, restore wetlands, overhaul city drainage systems and add more green-blue spaces to absorb water.

Hurricanes are measured using the Saffir-Simpson Hurricane Wind Scale, which categorizes them as 1 to 5 based on the wind speed produced by the hurricane. A Category 5 hurricane is the strongest, with wind speeds of 157 miles per hour (250 kilometers per hour) or higher.
WARREN FAIDLEY/GETTY IMAGES

Antarctica was officially discovered about 200 years ago. While no one permanently lives there, scientists and researchers stay on the continent for periods of time to study the climate and animals that call the region home.
PAUL SOUDERS/GETTY IMAGES

 Thwaites Glacier   Antarctica   2026

DOOMSDAY GLACIER

Antarctica holds 90 percent of Earth's ice. A huge part of it, known as Thwaites Glacier, sits in West Antarctica. It is huge—over half a mile (a kilometer) thick and about the size of Great Britain or Florida. In the last 10 years, the ice has shrunk by over 16 feet (4 meters). Warmer ocean water is eating away at the glacier. **Glaciologists** have warned that should the ice shelf collapse, it will raise the sea level by three meters.

The rapidly vanishing ice contributes to warming the planet. The ice acts like a giant umbrella, reflecting the sunlight and keeping the planet cool and its climate stable. As the ice melts, sunlight hits the darker ocean surfaces and is absorbed, which warms the region. That in turn melts more ice, creating further warming, and on it goes. This vicious cycle is part of the reason polar regions are warming at least two times faster than the rest of the planet.

WHAT DOES IT MEAN FOR HUMANITY?

Just like when the glaciers melted 11,000 years ago, rising sea levels will affect humans, especially the 470 to 760 million people living in low-lying coastal areas. Scientists warn of increased damage from hurricanes and other storms, eroding coastlines and storm surges. Small island nations like Kiribati, the Maldives and Tuvalu are at risk of disappearing altogether because of their low elevation.

WHAT CAN WE DO?

The good news is that we have the resources and knowledge to curb the climate crisis. In 2015, 195 countries adopted the Paris Agreement, agreeing to work toward keeping rising average global temperatures below 3.6°F (2°C) by 2030. Countries can invest in green alternatives like wind and solar energy, plant more trees and make homes and buildings more energy-efficient. *You* can help too! Grow a garden, ride your bicycle instead of taking a car, and turn off lights, TVs and computers when you aren't using them. These might seem like small things, but they add up. As anthropologist Margaret Mead said, "Never doubt that a small group of thoughtful, committed citizens can change the world; indeed, it's the only thing that ever has."

After experiencing Hurricane Sandy, Vic Barrett (left) grew concerned about US energy policies and their effect on the environment. In 2019 Barrett joined a lawsuit claiming that the federal government violated people's rights by allowing industries to put dangerous levels of carbon dioxide into the atmosphere.
REPRESENTATIVE BILL KEATING/WIKIMEDIA COMMONS/PUBLIC DOMAIN

EIGHT

EQUITY
MAKING WATER SAFE AND ACCESSIBLE FOR EVERYONE

In developed countries, water goes through a complicated treatment process before it is piped into homes. Even in a country like Canada, however, clean and safe drinking water is not a reality for all citizens. More than 50 Indigenous communities are under drinking-water advisories. Water equity sometimes becomes a political issue, which is why, in 2010, the United Nations declared access to clean water an essential human right. It is the job of governments to ensure that *all* citizens have access to clean water and proper sanitation.

Outbreaks of disease were commonplace in crowded 19th-century cities. Five hundred people died of cholera in the summer of 1854 in London because of a contaminated drinking well. Dr. John Snow traced the origins of the disease to a mother who had washed her baby's dirty diapers in the sink of her home, which then drained into the well. As soon as the well was shut down, the cholera epidemic ended, and better wastewater management began—at least in most of the developed world. Unsafe drinking water is a problem that still affects one in four people around the world. There are lots of reasons water isn't safe to drink.

- Poor sanitation
- **_Saltwater intrusion_**
- Industrial or agricultural pollution

ROMAN INGENUITY

Around the city of Rome, you'll find ancient drinking fountains called nasoni, so named because the singular *nasone* translates to "large nose," and the curved, downward-pointing metal spout of the fountains resembles a nose. Nasoni are part of the original aqueduct system dating to 312 BCE. The fountains are always flowing with fresh, drinkable water and are located all over the city. The added benefit is that the running taps keep the water in the pipes from getting stagnant.

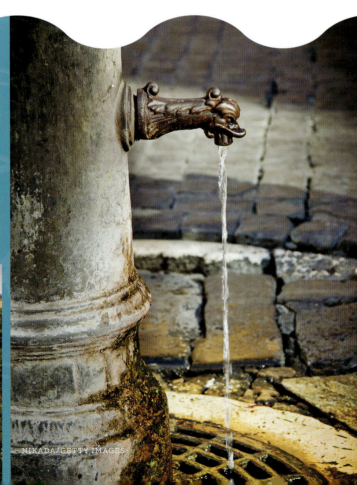

Shoal Lake 40, Ontario-Manitoba Border, Canada   North America   1919

DO YOU KNOW WHERE YOUR WATER COMES FROM?

It was exciting for Winnipeggers to have clean, clear water run from their taps in April 1919. A water-filtration plant and a 96-mile (155-kilometer) aqueduct from Shoal Lake, which straddles the Manitoba-Ontario border, had been completed to provide safe drinking water for the growing prairie town. But for the Anishinaabeg who live on Shoal Lake 40 First Nation and whose reserve also straddles the border, it was a different story. They didn't have access to the purified water that was being sent to Winnipeg. For the next seven decades, they drank unfiltered lake water.

WATER ADVISORY BEGINS FOR SHOAL LAKE

In 1997 a boil-water advisory was issued for Shoal Lake 40 because of an outbreak of cryptosporidiosis, a highly contagious intestinal infection. For the next 24 years, bottled water had to be hauled in by barge during the summer months and over ice roads in the winter. Generations of people have never known what it's like to drink water straight from the tap.

Shoal Lake 40 is not the only community in Canada to face these problems. In fact, there are hundreds of drinking-water advisories from "boil water" to "do not consume," and most of them are for Indigenous communities. After much media attention and a visit from the prime minister, Shoal Lake 40 reserve finally received a new water-treatment facility in 2021. For the first time in more than 20 years, people in the community were able to do what most Canadians take for granted—drink water straight from the tap.

Water-treatment plants ensure that the water coming out of your taps is clean and safe to drink. Iron, chloride, aluminum salts and other chemicals are added in a process called coagulation. Filtration removes bacteria and viruses that make people sick.
OGNIANM/GETTY IMAGES

LHEIDLI TOO (TWO RIVERS WATER)

Inspired by water activists like Autumn Peltier, actor, dancer and DJ Keilani Rose organized a group of students from University of Northern British Columbia (UNBC) to raise questions about the ongoing boil-water advisories in her home community, the Lheidli T'enneh reserve.

The reserve sits at the point where the Nechako River enters the Fraser River in British Columbia. Despite the abundance of fresh water, outdated infrastructure has resulted in boil-water advisories for drinking water on and off since 1913. Refusing to accept the conditions, Rose mobilized a group of people committed to improving access to water and quality of life on the reserve. Along with organizations like the Dream Catcher Foundation and Healthy First Nations, the Lheidli Too campaign is bringing awareness to the issue, calling on the Canadian government to help and providing temporary water-filtration units to locations in need.

MICHAEL BEZJIAN/GETTY IMAGES

SHUNYU FAN/GETTY IMAGES

Cox's Bazar, Bangladesh Asia 2021

WATER CRISIS IN REFUGEE CAMPS

Cox's Bazar in Bangladesh is a refugee camp inhabited by 900,000 stateless Rohingya people. With open sewage, seasonal flooding, rainstorms and lack of proper sanitation, waterborne illnesses like cholera, diarrhea and typhoid fever spread easily when people unknowingly drink, bathe, cook or play in contaminated water.

A similar crisis developed when the record-breaking rains of 2022 overwhelmed the banks of the Indus River in Pakistan. Floods wiped out homes, cattle, crops and livelihoods. Millions were left homeless and vulnerable to illness and starvation. Even with all that water, people were left thirsty because the heavy rains swept animal and human waste into the water source, making it unsafe to drink. Even the most advanced water-treatment facilities couldn't keep up.

Ensuring that almost a million people have clean water is a challenge. The UN's High Commissioner for Refugees (UNHCR) wants to provide people in refugee camps with about 5 gallons (20 liters) of water per person per day but estimates that more than half of the camps are not able to meet these requirements.
CAPTAIN RAJU/WIKIMEDIA COMMONS/CC BY-SA 4.0

WHAT ARE THE SOLUTIONS?

Ensuring that millions of people have clean water during an emergency is a challenge. Water can be trucked in at a great expense, or narrow openings called **boreholes** can be drilled so water can be extracted from the ground with a hand pump. The United Nations High Commissioner for Refugees has introduced solar-powered motorized pumps that draw water from chlorinated tanks and pipe it to taps installed throughout Cox's Bazar.

Hamza Farrukh is an ecopreneur and founder of the charity Bondh E Shams. He developed a solution to help in an emergency water crisis. The Oasis Box uses solar energy to filter floodwater into clean, drinkable water, and it has a storage tank. Farrukh knows firsthand how important clean water is. When he was seven, he contracted typhoid in his village after drinking contaminated water. He believes that innovations like the Oasis Box could help end the global water crisis by 2050.

SOLUTIONS TO CHANGE THE WORLD

The Stockholm Junior Water Prize is awarded annually to a student between the ages of 15 and 20 who has developed a research project that can help solve a major water challenge, like equity or access, at a global or local level. Projects are reviewed by a jury of international water experts and a winner is selected. Past winners include Eshani Jha, who won for her research on how to remove contaminants from water, and Rachel Chang and Ryan Thorpe, who created a way to detect and purify water contaminated with harmful bacteria like *E. coli* and *Vibrio cholerae* (cholera). As Jha said in her acceptance speech, "We really are the future of water-related science." She's right! A generation of young people motivated to solve society's biggest water challenges can and will change the world.

📍 California, US   🌐 North America   🕒 2035

H2O IN THE OC

In 1975 the people of Orange County, California, had a problem. Not only had a lack of precipitation and overconsumption of water depleted aquifers, but seawater had seeped into them, making what little water remained undrinkable. Action needed to be taken. With the help of **water engineers**, Orange County Water District designed a wastewater-to-drinking-water treatment plant.

TOILET TO TAP

While it might not sound appetizing, researchers have found that "toilet to tap," or recycled wastewater, is as clean, if not cleaner, than surface-water sources that supply many cities. Because of the bias toward drinking wastewater, it is often put through additional rigorous testing and treatment, making it some of the safest you can drink. Today the system delivers high-quality drinking water for a million people, making the project a success *and* the largest facility of its kind in the world. Innovative, sustainable strategies like wastewater recycling are needed in regions like California, which is expected to lose 10 percent of its water supply by 2035 due to hotter, drier weather.

Following Orange County's lead, Los Angeles has included fully reusing its water supply in its long-term water management objectives by building an advanced water purification facility (AWPF) to treat and purify up to 15 million gallons (68 million liters) of wastewater per day.
AVAILABLELIGHT/GETTY IMAGES

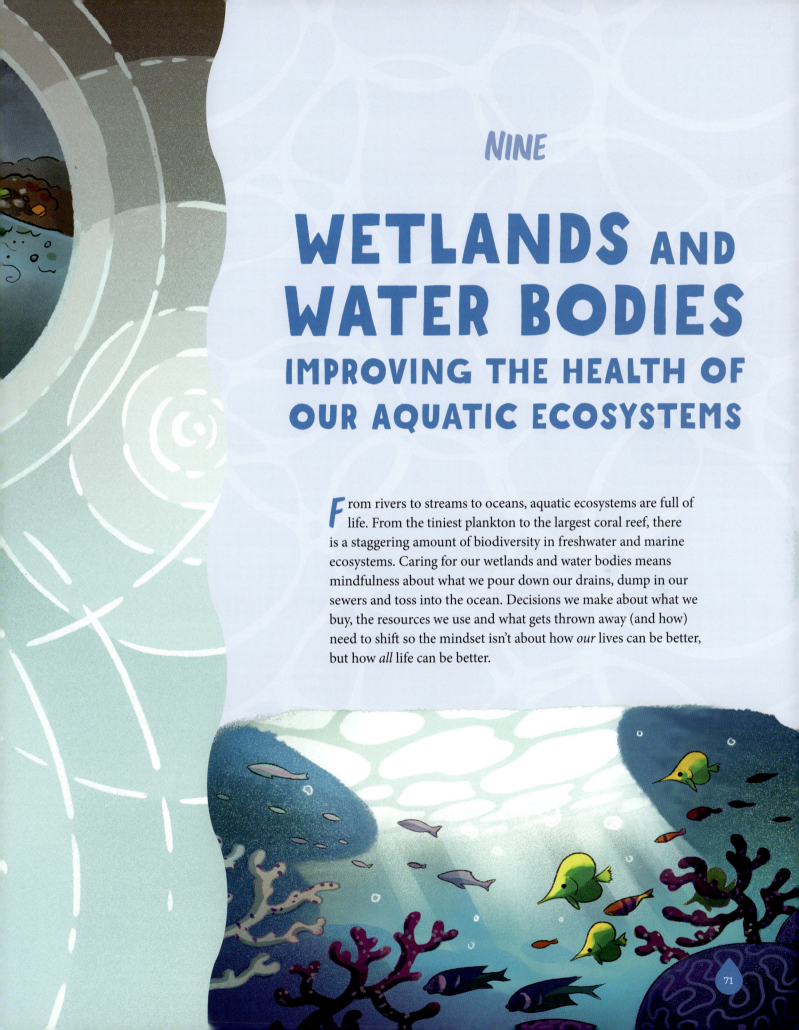

NINE

WETLANDS AND WATER BODIES
IMPROVING THE HEALTH OF OUR AQUATIC ECOSYSTEMS

From rivers to streams to oceans, aquatic ecosystems are full of life. From the tiniest plankton to the largest coral reef, there is a staggering amount of biodiversity in freshwater and marine ecosystems. Caring for our wetlands and water bodies means mindfulness about what we pour down our drains, dump in our sewers and toss into the ocean. Decisions we make about what we buy, the resources we use and what gets thrown away (and how) need to shift so the mindset isn't about how *our* lives can be better, but how *all* life can be better.

Chesapeake Bay, Maryland, US   **North America**   **1983**

Of the 32 largest cities in the world, 22 are located on estuaries.
JACQUES DESCLOITRES, MODIS LAND SCIENCE TEAM/NASA

WHAT'S GOING ON?

In the 1970s people who lived and worked in the Chesapeake Bay *estuary* noticed algae was growing out of control. For years residential waste and agricultural runoff from chemicals used in farming had been entering the watershed. Excess nitrogen and phosphorus encouraged the growth of algae, a process called eutrophication. Too much algae in an aquatic ecosystem leads to hypoxia—low or depleted oxygen levels. Hypoxic conditions create dead zones, areas where the oxygen is too low to support marine life. In an estuary like Chesapeake Bay, where fish, crabs and oysters lived, that was bad news.

RESTORATION BEGINS

Recognizing the work that needed to be done to repair Chesapeake Bay, governments acted together and signed the Chesapeake Bay Agreement in 1983. It was the first estuary in the United States targeted for restoration—a signal that people understood the value of these ecosystems and the negative impact humans could have on them. The Chesapeake Bay Foundation has planted trees as buffers along rivers and streams, controlled pollution and raised awareness through education of the importance of keeping waterways free of dangerous chemicals. Chesapeake Bay isn't out of danger. Pollution continues to be a problem, but a collective effort from citizens and the government will protect the estuary for years to come.

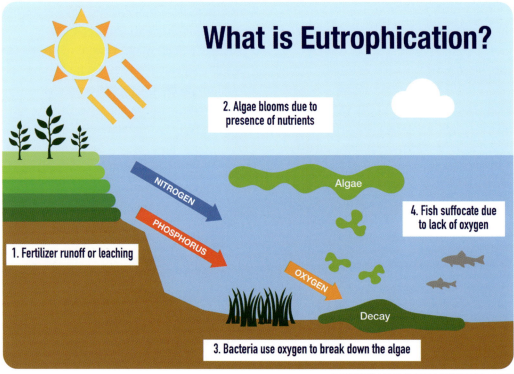

DIMITRIOS KARAMITROS/GETTY IMAGES

📍 Prince William Sound, Alaska, US 🌐 North America 🕑 1989

ENVIRONMENTAL DISASTER

The pristine wilderness of Prince William Sound, an inlet in the Gulf of Alaska, was destroyed when an oil tanker, *Exxon Valdez*, ran into a cold-water reef and spilled nearly 11 million gallons (50 million liters) of crude oil. The spill not only polluted the water but also killed many animals, including salmon, sea otters, bald eagles and orcas.

Cleaning up oil spills is labor-intensive. Boat-based skimmers are used to collect the oil and burn it, which further contaminates the air. Chemicals are used to break up the oil into smaller droplets. Animals coated in oil need to be captured and washed with gentle dish soap before being released.

LINGERING EFFECTS

Despite people's best efforts, oil spills can't be fully cleaned up. Even decades later, the effects of the oil are being felt in coastal and fishing communities and the ecosystem. The salmon and herring populations are still recovering, as are local populations of orcas and some seabirds. The emotional impact of seeing unspoiled nature destroyed made people rethink how we use our waterways and what we need to do to protect them. After the spill, US Congress passed the Oil Pollution Act, which created procedures for responding to future oil spills. The act also required all new tankers to be built with double-walled hulls to guard against spills and phased out the use of single-hulled tankers like *Exxon Valdez*.

The Exxon Valdez oil spill killed an estimated 250,000 sea birds, 3,000 otters, 300 harbor seals, 250 bald eagles and 22 orcas. Exxon employees, federal responders and more than 11,000 Alaskan residents worked to clean up the oil spill.
MIKE SHOOTER/SHUTTERSTOCK.COM

DRILLING AND DESTRUCTION

Emergildo Criollo, leader of the Kofan People and cofounder of Ceibo Alliance, lived by the Amazon River. At six years old, while hunting and fishing with his family, he heard a loud noise in the sky. Thinking it was some kind of large bird, Criollo hid. It was actually a helicopter delivering workers to cut down trees for an oil-drilling operation. Within a few months, great waves of oil floated down the river. Criollo and his family continued to fish and drink the water, not realizing how polluted it had become. As Criollo grew older, the polluted water made his people sick with oil-related illnesses and ravaged the region he called home. Along with members of other Indigenous communities, Criollo formed Ceibo Alliance, with the goal of installing rainwater harvesting systems so people would have access to unpolluted water.

Criollo and the other members of the alliance are working together to fight for oil companies to take responsibility for the damage they have caused and ensure that the Amazon River is protected for future generations.

As demand for oil and its price continues to climb, oil companies are willing to go anywhere—even to the middle of the Amazon Rainforest—to set up drilling operations.
PILESASMILES/GETTY IMAGES

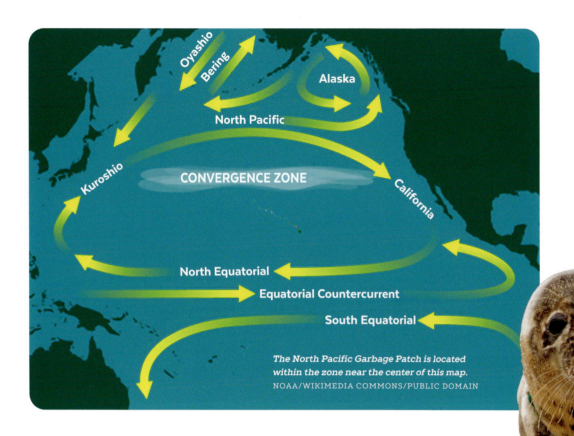

The North Pacific Garbage Patch is located within the zone near the center of this map.
NOAA/WIKIMEDIA COMMONS/PUBLIC DOMAIN

📍 North Pacific Garbage Patch   🌐 Pacific Ocean   🕒 1997

FLOATING GARBAGE DUMP

In 1997 racing-boat captain Charles Moore was crossing the Pacific Ocean through what should have been unspoiled ocean when he noticed that all he could see was plastic litter. He was the first person to identify the North Pacific Garbage Patch.

Garbage patches are created when gyres, or rotating ocean currents, pull floating debris into one location. It is hard to know how much garbage is in the garbage patch because it stretches from the surface down to the bottom of the seafloor, moving with the winds and currents.

It is estimated that 80 percent of the plastic in oceans comes from land-based sources. Bottle caps, water bottles and plastic bags are examples of single-use plastics discarded by humans that find their way into rivers and spill into the oceans. The other 20 percent comes from boats, including discarded fishing nets left to drift. Plastic is nonbiodegradable, which means it will break down into smaller and smaller pieces but will never completely disintegrate.

IMPACT ON THE OCEANS

The ocean plastic adversely affects marine life. Turtles eat plastic bags thinking they are jellyfish, and albatross mistake plastic pellets for eggs and feed them to their babies. Seals, dolphins and whales get trapped in discarded fishing nets and die. And the plastic floating on the surface blocks sunlight, stopping the growth of the algae and plankton that the entire marine food web is based on.

Abandoned fishing nets, or "ghost nets," pose a danger to marine animals, which can get trapped or tangled in the gear. The World Wildlife Federation (WWF) would like to see stricter rules put in place to make sure damaged nets are not lost or left behind.
KEV GREGORY/SHUTTERSTOCK.COM

WHAT CAN WE DO?

The first and most important step in dealing with the garbage patch is to create less garbage. According to the UN, 295 million tons (300 million metric tons) of plastic waste is produced annually, and only 9 percent is recycled. Over 16 million tons (17 million metric tons) of plastic ends up in oceans each year. Beach cleanups, sea bins that filter garbage out of waterways, and government efforts to stop trash from entering the oceans will reduce plastics, but the very best way to keep the oceans clean starts with you. Recycling, making smart buying decisions and eliminating single-use plastics are simple and easy ways to help.

Participants in the Yellow Fish Road program educated community members about protecting waterways by painting yellow fish symbols with the words "Rainwater Only" by storm drains.
KEVIN M. LAW/ALAMY STOCK PHOTO

According to the Interceptor's inventor, Boyan Slat, rivers are "the arteries that carry the trash from land to sea." He was 18 years old when he founded the Ocean Cleanup to develop and deploy the system he invented.
THE OCEAN CLEANUP

TRAPPING THE TRASH

With the goal of removing 90 percent of floating ocean plastic by 2040, the Ocean Cleanup uses a mobile U-shaped barrier to concentrate and guide plastic into a retention zone so it can be extracted. It's a large-scale, energy-efficient system that is used in every ocean gyre. The company also invented a solar-powered plastic extractor for rivers so trash can be trapped before it gets to the ocean. The Interceptor Original runs 24 hours a day, 7 days a week.

Once waste is collected from waterways, what do you do with it? Dutch company Clear Rivers repurposes plastic trash collected in a litter trap into a series of hexagonal floating pods that can be used as rest areas or gardens.

CLEAR RIVERS

The Seal River Watershed is one of the largest watersheds in the world, covering 19,000 square miles (50,000 square kilometers), and it has very little development like roads or mines.
STUART FORSTER/ROBERTHARDING/GETTY IMAGES

| Seal River Watershed, Manitoba, Canada | North America | 2030 |

PROTECTING OUR WATERSHEDS

A group of Indigenous Guardians made an offering of tobacco before they set off on a seven-day canoe journey up the Seal River to Hudson Bay in August 2022. The Seal River Watershed in northern Manitoba spans an area the size of Nova Scotia and is home to eagles, beluga whales, polar bears, moose and caribou. It is the ancestral territory of numerous Indigenous nations and communities, including the Sayisi Dene First Nation, Northlands Denesuline First Nation, Barren Lands First Nation and O-Pipon-Na-Piwin Cree Nation. These four nations created the Seal River Watershed Alliance with the goal of making the watershed an *Indigenous Protected and Conserved Area (IPCA)*.

BLENDING SCIENCE WITH WISDOM

Indigenous Guardians, people who care for traditional territories on behalf of their individual nations, play a key role in the Seal River Watershed Alliance's conservation strategy. Guardians are trained experts who blend ancestral knowledge of land from Elders with western science practices like water sampling, trail cameras and drones. There are about 120 Indigenous Guardians programs operating across Canada.

30 BY 30

As biodiversity loss and the climate crisis alter our world, governments are realizing how important it is to conserve and protect land. At the 15th Conference of the Parties (COP15), a target was set by world leaders to protect 30 percent of the world's lands and oceans by 2030, in hopes of curbing biodiversity loss and limiting the damaging impacts of the climate crisis. Manitoba, a province in the middle of Canada, has only 11 percent of its land and water protected. In 2024, the Seal River Watershed Alliance secured interim protection for the watershed from the provincial and federal governments. It's the first step to protecting the watershed for the future and moving the province closer to the COP15 goal of 30 by 30.

SWELL OF HOPE

You may have heard of the 4 Rs: recycle, reuse, reduce and repurpose. When it comes to caring for and conserving water, there's a fifth: relationship. Understanding how we impact water in our daily lives and how important water is for our survival is the first step to ensuring we treat it with the respect it deserves. As one scientist I spoke to said, "We need water, but it doesn't need us."

Through programs at schools, camps and nature centers, you can learn more about what you can do to protect our most important natural resource. Getting outdoors to enjoy the beauty of oceans, lakes and rivers near you will foster your desire to keep them healthy and inspire others to do the same.

Water is a powerful force, one that has shaped our world and the people who call it home. Your actions, no matter how small, are what will create a better future for our blue planet.

Think about the way water connects all living things, from plants to animals to humans, the next time you take a drink, jump in a lake or feel the rain on your face.
FEI YANG/GETTY IMAGES

The United Nations designated March 22 as World Water Day. Around the world events are organized to raise awareness about water conservation, equity and accessibility. But you don't have to wait for March 22. Our world's water deserves to be celebrated and cherished every day!
RONNIECHUA/GETTY IMAGES

GLOSSARY

agroforestry—growing crops along with trees and shrubs

aquifer—an underground layer of permeable rock that contains water and sometimes gravel, sand or silt

biodiverse—having a large variety of plants and animals

boreholes—narrow openings, often drilled by hand, for extracting water

clean energy—energy that doesn't rely on the burning of fossil fuels like coal and oil

currents—continuous movement of lake or ocean water, created by forces such as wind and temperature. Currents are like vast underwater rivers that move warming and cooling water.

deep-sea mining—removing minerals from a seabed 656 feet (200 meters) or more below the surface

delta—landform shaped like a triangle at the mouth of a river

dykes—embankments that control or hold back the water of a river or the sea

estuary—a water passage where salty tidal waters meet a freshwater river

feedback loop—a chain of cause and effect that forms a circuit, or loop. In the climate system, some feedback loops accelerate temperature rise, others decelerate it.

fossil fuels—natural substances that can be used as a source of energy

glaciers—large sheets of ice

glaciologists—scientists who study and analyze the physical properties and movement of ice

gray water—wastewater from showers, baths, washing machines and kitchen sinks, which typically doesn't contain contaminants (not wastewater from toilets)

groundwater—water present beneath the earth's surface

hydroponic—referring to the process of growing plants without soil

Indigenous Protected and Conserved Area (ICPA)—an area of lands and waters for which Indigenous governments are responsible for protecting and conserving its ecosystems

Industrial Revolution—the period, beginning in the 18th century, when there was a shift from an agricultural society to one dominated by industry and machine manufacturing, often relying on coal for power

infrastructure—the systems and services, including buildings, roads and power supplies, that enable a city, region or country to function effectively

internally displaced person—someone forced to move within their country due to climate-related issues like floods and droughts

irrigation—the watering of land by artificial means

metropolis—a large city

net-zero emissions—the state where a country removes as much carbon dioxide from the atmosphere as it releases into it, reducing emissions and offsetting them with positive actions like tree planting

oceanographer—a scientist who studies the ocean's chemistry, geology and biology

renewable energy—energy which comes from replenishable sources like the sun, wind and tides

reservoirs—large natural or artificial lakes where water is collected and stored

saltwater intrusion—the movement of saline water into freshwater aquifers

sonar—an acronym for sound navigation and ranging, a technique that uses sound waves to detect objects underwater and measure the water's depth

standard of living—the level of wealth and material comfort of the average person in a community

submersible—an underwater vehicle

trade winds—winds that blow steadily near the equator, from the northeast in the northern hemisphere and from the southeast in the southern hemisphere, especially at sea. Two belts of trade winds encircle the earth, blowing from the tropical high-pressure belts to the low-pressure zone at the equator.

water engineers—people who design, construct and plan the maintenance of systems that collect, clean and distribute water. Projects often involve groundwater, hydraulics, pumping systems and dams.

water footprint—a measure of the amount of water used to produce each of the goods and services we use

watershed—an area of land that drains, or "sheds," rainfall and snowmelt into a body of water such as streams and rivers

water-stressed—a term to describe a region that withdraws 25 percent or more of its renewable freshwater resources

water vulnerability—at risk or in danger of losing access to clean water

RESOURCES

Books

Castaldo, Nancy F. *When the World Runs Dry: Earth's Water in Crisis.* Algonquin Young Readers, 2022.

Curtis, Andrea. *City of Water.* Groundwood Books, 2021.

Mulder, Michelle. *Every Last Drop: Bringing Clean Water Home.* Orca Books, 2022.

Rae, Rowena. *Upstream, Downstream: Exploring Watershed Connections.* Orca Books, 2021.

Strauss, Rochelle. *One Well: The Story of Water on Earth.* Kids Can Press, 2007.

Strauss, Rochelle. *The Global Ocean.* Kids Can Press, 2022.

Online

Equity and security: water.org

Weather and climate change: climatekids.nasa.gov

Weather, climate change and wetlands: noaa.gov

Protecting oceans: missionblue.org

Water use: waterfootprint.org

Water cycle: kids.nationalgeographic.com

Conservation: swimdrinkfish.ca

Watersheds and conservation: sealriverwatershed.ca

World Water Day: unicef.ca/en/blog/celebrate-world-water-day-children-around-world

Water Crisis: thewaterproject.org

Video

Water Fix!: Crash Course Kids #36.2 and **Water Fight!**: Crash Course Kids #36.1, Crash Course Kids YouTube channel

Links to external resources are for personal and/or educational use only and are provided in good faith without any express or implied warranty. There is no guarantee given as to the accuracy or currency of any individual item. The author and publisher provide links as a service to readers. This does not imply any endorsement by the author or publisher of any of the content accessed through these links.

ACKNOWLEDGMENTS

A huge thank-you to the folks at Orca, especially to my editor, Kirstie Hudson, who guided me through the process of writing this book. The incredible illustrations are thanks to Sophie Dubé. Much gratitude also goes to editorial assistant Georgia Bradburne, copyeditor Vivian Sinclair and designer Dahlia Yuen for keeping things organized and making the book look so good.

    The idea for this book originated with an inquiry project my teaching partner, Alex McGavin, and I developed for our class. We asked the students to decide which had the bigger impact, water on humans or humans on water. The question led to many spirited discussions and was the catalyst for a deeper dive into how to bring our idea alive in a book.

INDEX

*Page numbers in **bold** indicate an image caption.*

activists
   ocean cleanup, 77
   research projects, 52–53, 68
   and stewardship, 45, 78–79
   use of lawsuits, **61**
   water as human right, 33, 35, 66
   See also solutions
Aden, Yemen, 7
agriculture
   beef production, 13
   and drought, 17–19, 30, 57
   and fresh water use, 13–14, 17–19
   hydroponics, **12**, 19, 81
   irrigation, 14–16, 18–19, 82
   loss of farmland, 19
   runoff water, 72
   seasonal flooding, 6, 15
   and waterwheels, 22
agroforestry, **12**, 19, 81
algae blooms, 72
Amazon River, **46**, 50–51, 74
Antarctica, 53, **54**, 60
aquifers
   defined, 81
   and groundwater, 17–19, 35, 81
   and overuse, 35, 36
   saltwater intrusion, 10, 64, 82
atmosphere
   hurricanes, **11**, 58–59
   trade winds, 82
Aztec farming practices, 15

Badiel, Georgie, 33
Bajau People, **44**
Barbegal watermill, **20**, 22
Barlow, Maude, 35
Barrett, Vic, **61**
beef production, 13
Bering Strait, **54**, 56
beverage industry, 35
biodiversity, 51, 71, 81
Blue Marble Project, **38**, 45
boreholes, 68, 81
Buddhism, 42

Busan, South Korea, **11**

California, 69
Canada
   daily water usage, 32
   drought in, 17
   ownership of water, 35
   and safe drinking water, 65–66
Cape Town, South Africa, 34
Chesapeake Bay estuary, 72
cholera outbreak, 64
Christians, 42
cities
   floating, **11**
   and flooding, **4**, 10, 58–59, 60, 68
   lakefront restoration, 45
   as ports, **5**, 6–9
   wastewater, 72, 76
   water retention, 36, **37**
   water shortages, 34, 36
Clarion-Clipperton Zone, 52
clean energy, 25, 81
   See also renewable energy
climate crisis
   clean energy, 24, 25, 81
   feedback loop effect, 30, 81
   fossil fuels, 25, 55, 56
   and global warming, 55–61
   and internally displaced persons, 10, 81
   international agreements, 24, 56, 61, 79
   lawsuits, **61**
climate justice
   access to water, 33, 63–69
   daily water usage map, 32
   internally displaced persons, 10
   ownership of water, 35
   water as human right, 33, 35, 66
   water-stressed regions, 36, 82
Colorado River Basin, 57
consumer choice. *See* solutions
Cox's Bazar, Bangladesh, **62**, 67–68
Criollo, Emergildo, 74

currents
   defined, 81
   ocean, 55, 57
   and tidal energy, 27

dams, 21, 24, 25–26
Day Zero, **28**, 34
deep-sea mining, 52, 81
deltas, 15, 81
desalination plants, **28**, 31
desertification, 19
Dhaka, Bangladesh, 10
diseases, waterborne, 64–68
drinking water
   access to, 33, 63–69
   fountains, **62**, 64
   from recycled wastewater, 69
   unsafe, 63–68, 74
drought
   and agriculture, 17–19, 30, 57
   and cities, 34, 36
dykes, 58, 81

Earle, Sylvia, 52, 55
ecosystems
   aquatic, 71–76
   biodiverse, 51, 71, 81
Egyptians, ancient, 15
Ellis Island, NYC, **4**, 8
estuaries, 72, 81
exploration
   boats and ships, 47–51
   underwater, 52–53
Exxon Valdez disaster, **70**, 73

farming. *See* agriculture
Farrukh, Hamza, 68
feedback loop, 30, 81
flooding
   flash floods, 10, **54**, 57
   and rising sea levels, 55, 56, 58, 60
   seasonal, 6, 15
   storm surges, 10, 58, 59, 61

food security
    farming practices, 18–19
    and hydroponics, **12**, 19, 81
    local produce, 15
fossil fuels
    defined, 81
    oil industry, 73, 74
    use of, 25, 55, 56
freshwater
    lakes, 1, 57, 65
    wetlands, 17, 18, **37**, 59
    See also rivers

Ganga River, India, **38**, 41
gardening, **14**, 15
gender inequality, 33
glaciers, **54**, 55, 56, 60
global warming, 55–61
governments
    international agreements, 24, 56, 61, 79
    lawsuits against, **61**
    net-zero emission goal, 56, 82
    ownership of water, 35
    role of, 29, 43, 58–59
    and safe drinking water, 63, 65–66
Grand Canal, China, **20**, 23
gray water, 34, 81
groundwater, 17–19, 35, 81

habitat restoration
    coastal, **56**, 72
        garbage cleanup, 75–77
        lakefront, 45
        and oil spills, 73, 74
        river rights, 43
        watersheds, 57, 78–79
Hawaiians, ancient, **38**, 40
Heinerth, Jill, 53
Hinduism, 41
Huang He (Yellow River), **6**, 23
Hurricane Fiona, **54**, 58
hurricanes, **11**, 58–59
hydroelectric power, **20**, 21, 24, 25–26
hydroponics, **12**, 19, 81

Incan terraced farming, 16
Indigenous Guardians programs, 78–79

Indigenous Peoples
    and safe drinking water, 63, 65–66
    settlements, 9
    spirituality, 40, 43, **44**
    and stewardship, 26, 74, 78–79
Indigenous Protected and Conserved Area (IPCA), 78–79, 81
Indus River, **6**, 67
Industrial Revolution, 9, 23, 81
infrastructure, 10, **58**, 81
innovation
    desalination plants, **28**, 31
    emergency water supply, 68
    floating cities, **11**
    garbage cleanup, **70**, 77
    hydroponics, **12**, 19, 81
    submersibles, 53, 82
international agreements, 24, 56, 61, 79
irrigation, 14–16, 18–19, 82

Jews, 42
Jha, Eshani, 68

Kyoto, Japan, 40

lakefront restoration, 45
lakes, 1, 57, 65
Lheidli T'enneh reserve, BC, **62**, 66
London, cholera outbreak, 64

Magellan's voyages, **46**, 49
Māori tribes, 43
Mattson, Mark, 45
Mayans, **28**, 30
Mead, Margaret, 61
Mehgna River, Bangladesh, 10
Mesopotamia, **4**, 6
metropolis, 6, 82
Mexico City, 15, 36
Mississippi River, **4**, 9
Muskrat Falls Generating Station, Canada, 26
Muslims, 42

natural resources
    mining, 52
    oil industry, 73, 74

nature
    respect, 1–**3**
    rights of, 43
    and spirituality, 39–44
    and wellbeing, 45
Nestlé, 35
Netherlands, 58
net-zero emissions, 56, 82
New Orleans, LA, 9
New York City, NY
    hydroponics in, 19
    port city, **4**, 8
Nile River delta, **6**, **12**, 15
North Pacific Garbage Patch, **70**, 75–76

Oasis Box project, 68
Ocean Cleanup project, **70**, 77
oceans
    and coastal cities, 10, 60
    coastal habitat, **56**
    dead zones, 72
    deep-sea mining, 52
    floating garbage, 75–77
    and global warming, 55–56, 58–61
    and Indigenous cultures, 40, **44**
    seafloor mapping, **46**, 52, 53
    and trade routes, 7–8, 47–49
oil spills, **70**, 73
Orellana, Francisco de, **46**, 50

Pakistan, flooding, 67
Philippine Cordilleras, 16
plastic waste, 35, 75–77
precipitation. *See* rainfall
Prince William Sound, AK, 73
protected areas, 52, 78–79
public awareness, 1–2, 35, 45, 76, 79

rainfall
    and climate patterns, 57
    collection, 2, **36**
    feedback loop effect, 30, 81
Ras Al-Khair Power and Desalination Plant, Saudi Arabia, **28**, 31
refugee camps, **62**, 67–68

renewable energy
   defined, 82
   hydroelectric power, **20**, 21, 24, 25–26
   and natural resources, 52
   tidal energy, **20**, 27
reservoirs, 17, 18, 57, 82
resources, 45, 83
rivers
   and biodiversity, 51, 81
   estuaries, 72, 81
   impact of dams, 21, 25–26
   legal identity, 43
   protected area, 78–79
   restoration of, 41, 43, 74
   sacred, 40, 41, 43
   seasonal flooding, 6, 15
   and trade routes, 6, 9–10, 50–51
   watershed, 57
Roberts, Tara, 53
Rome, ancient, **62**, 64
Rose, Keilani, 66

sacred waters, 40, 41, 43
saltwater intrusion, 10, 64, 82
Seabed 2030 Project, 52, 53
sea level, rise of, 55, 56, 58, 60
Seal River Watershed, MB, **70**, 78–79
Shanghai, China, 7
Shoal Lake 40 First Nation, 65
slave trade, 9, 53
solutions
   education, 76, 79
   government actions, 29, 43, 58–59
   innovative, 68
   international agreements, 24, 56, 61, 79
   less waste, 76
   and personal choices, 2, 29, 34, 61
   public awareness, 1–2, 35, 45, 76, 79
   urban design, 36, **37**
   use less water, 2, 29, 32, 34, 79
   See also habitat restoration
spirituality, 39–44
standard of living, 6, 82
Statue of Liberty, **8**

steam power, 9, **20**, 21, 23
Stockholm Junior Water Prize, 68
submersibles, 53, 82
surfing, **38**, 40
Swim Drink Fish organization, 45

terraced farming, **12**, 16
Three Gorges Dam, China, 25
Thwaites Glacier, **54**, 60
tidal energy, **20**, 27
Tokyo, Japan, 5
Toronto, ON, 45
trade routes
   and exploration, 47–51
   and industry, 9
   and port cities, **5**, 6–9
trade winds, 57, 82
transportation
   boats and ships, 7, 9, 47–51
   Grand Canal, China, 23
   oil tankers, 73
   steam power, 9, 23

United States
   climate change lawsuit, **61**
   coastal cities, 10
   daily water usage, 32
   immigrants' arrival, 8
   ownership of water, 35
   restoration projects, 72
urban. *See* cities
Uruk, 6

Vescovo, Victor, 53
Viking exploration, **46**, 48

wastewater
   impact of, 72, 76
   management, 64, 67
   recycled, 69
   water engineers, 69, 82
   water treatment plants, 65
water equity, 33, 63–69
water footprint
   daily water usage map, 32
   defined, 14, 82
water fuel cell hoax, 27
Watermark Project, 45

water pollution
   and algae blooms, 72
   cleanup, 41, 71–76
   and disease, 63–68
   floating garbage, 75–77
   and oil spills, **70**, 73
watersheds, 57, 78–79
water-stressed regions, 36, 82
water vulnerability
   in cities, 34, 36
   defined, 29, 82
   and drought, 17, 19, 30
   rural communities, 33
   world map, 32
   See also drinking water; innovation
waterwheels, 22
weather events, severe
   and drinking water, 67, 68
   and global warming, 55–59
   hurricanes, **11**, 58–59
   and infrastructure, 10, 81
   storm surges, 10, 58, 59, 61
wells, **28**, 33
wetlands, 17, 18, **37**, 59
Whanganui River, New Zealand, **38**, 43
wildlife
   and dead zones, 72
   floating garbage, 75–77
   impact of dams, 26
   and oil spills, 73
   and tidal turbines, **27**
Winnipeg, MB, 9
World Water Day, 80, 83

youth
   activists, 66, 68
   projects for, 45, 68

From the PAST to the PRESENT and into the FUTURE!

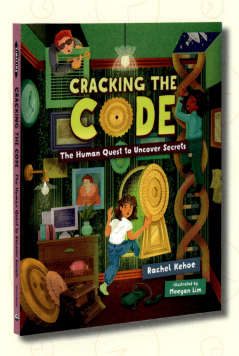

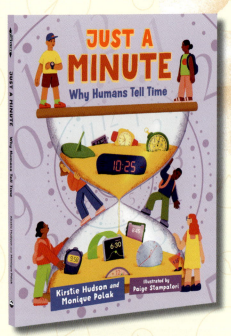

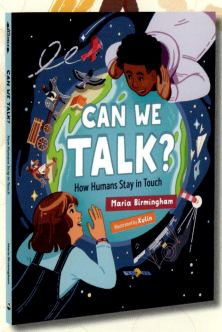

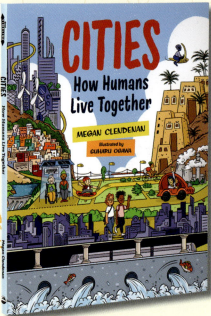

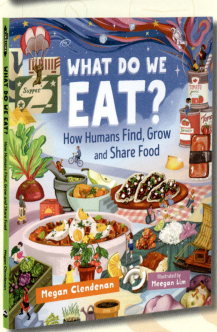

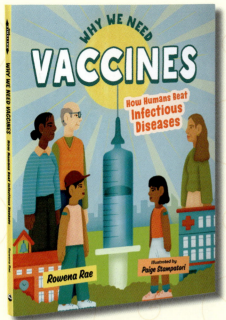  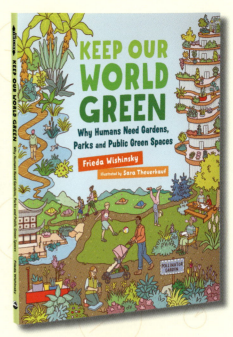

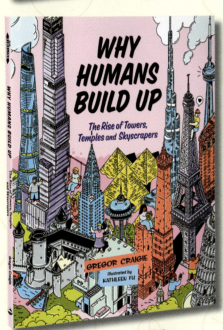 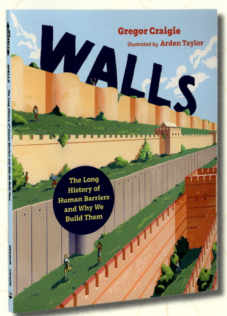

UPCOMING TOPICS

School
Engineering
Libraries
Sports

The Orca Timeline series explores how big ideas have shaped humanity. Discover what our collective history can tell us about the planet today and tomorrow.

Colleen Nelson is a teacher and award-winning author from Winnipeg, Manitoba. Her nonfiction titles include *If You Can Dream It, You Can Do It: How 25 Inspiring Individuals Found Their Dream Jobs* and *See It, Dream It, Do It: How 25 People Just like You Found Their Dream Jobs,* both co-written with Kathie MacIsaac. Her middle-grade novels include *The Umbrella House, Undercover Book List* and The Harvey Stories series. When not writing, Colleen loves spending time outside and globe-trotting, preferably to places near a lake, river or ocean.

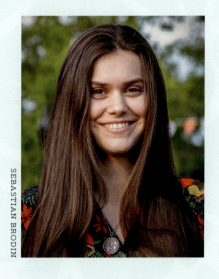

SEBASTIAN BRODIN

Sophie Dubé is a Canadian illustrator and concept artist. She graduated with honors from the Sheridan College illustration program and has worked as an art director for Zazie Films. Sophie lives in Vancouver, where she enjoys taking her sketchbook along for hikes and camping trips.